Representing Human Behavior in Military Simulations

Interim Report

Richard W. Pew and Anne S. Mavor, editors

Panel on Modeling Human Behavior and Command Decision Making:
Representations for Military Simulations

Committee on Human Factors
Commission on Behavioral and Social Sciences and Education
National Research Council

NATIONAL ACADEMY PRESS
Washington, D.C. 1997

NOTICE: The project that is the subject of this report was approved by the Governing Board of the National Research Council, whose members are drawn from the councils of the National Academy of Sciences, the National Academy of Engineering, and the Institute of Medicine. The members of the panel responsible for the report were chosen for their special competences and with regard for appropriate balance.

This report has been reviewed by a group other than the authors according to procedures approved by a Report Review Committee consisting of members of the National Academy of Sciences, the National Academy of Engineering, and the Institute of Medicine.

The National Academy of Sciences is a private, nonprofit, self-perpetuating society of distinguished scholars engaged in scientific and engineering research, dedicated to the furtherance of science and technology and to their use for the general welfare. Upon the authority of the charter granted to it by the Congress of 1863, the Academy has a mandate that requires it to advise the federal government on scientific and technical matters. Dr. Bruce M. Alberts is president of the National Academy of Sciences.

The National Academy of Engineering was established in 1964, under the charter of the National Academy of Sciences, as a parallel organization of outstanding engineers. It is autonomous in its administration and in the selection of its members, sharing with the National Academy of Sciences the responsibility for advising the federal government. The National Academy of Engineering also sponsors engineering programs aimed at meeting national needs, encourages education and research, and recognizes the superior achievements of engineers. Dr. William A. Wulf is interim president of the National Academy of Engineering.

The Institute of Medicine was established in 1970 by the National Academy of Sciences to secure the services of eminent members of appropriate professions in the examination of policy matters pertaining to the health of the public. The Institute acts under the responsibility given to the National Academy of Sciences by its congressional charter to be an adviser to the federal government and, upon its own initiative, to identify issues of medical care, research, and education. Dr. Kenneth I. Shine is president of the Institute of Medicine.

The National Research Council was organized by the National Academy of Sciences in 1916 to associate the broad community of science and technology with the Academy's purposes of furthering knowledge and advising the federal government. Functioning in accordance with general policies determined by the Academy, the Council has become the principal operating agency of both the National Academy of Sciences and the National Academy of Engineering in providing services to the government, the public, and the scientific and engineering communities. Dr. Bruce M. Alberts and Dr. William A. Wulf are chairman and interim vice chairman, respectively, of the National Research Council.

This work is sponsored by the Defense Modeling and Simulation Office, U.S. Department of Defense, and funded under contract DACW61-96-D-0001. The views, opinions, and findings contained in this report are those of the author(s) and should not be construed as an official Department of Defense position, policy, or decision, unless so designated by other official documentation.

Additional copies of this report are available from:

National Academy Press, 2101 Constitution Avenue, Box 285, Washington, DC 20418
800-624-6242 or 202-334-3313 (in the Washington metropolitan area).
http://www.nap.edu

Printed in the United States of America

PANEL ON MODELING HUMAN BEHAVIOR AND COMMAND DECISION MAKING: REPRESENTATIONS FOR MILITARY SIMULATIONS

RICHARD W. PEW (*Chair*), BBN Systems and Technologies, Cambridge, MA
JEROME R. BUSEMEYER, Department of Psychology, Purdue University
KATHLEEN CARLEY, Department of Social and Decision Sciences, Carnegie Mellon University
TERRY CONNOLLY, Department of Management and Policy, College of Business and Public Administration, University of Arizona, Tucson
JOHN R. CORSON, Integrated Visual Learning, Williamsburg, VA
KENNETH H. FUNK II, Industrial and Manufacturing Engineering, Oregon State University, Corvallis
DAVID E. KIERAS, Artificial Intelligence Laboratory, University of Michigan, Ann Arbor
RICHARD M. SHIFFRIN, Psychology Department, Indiana University, Bloomington
GREG L. ZACHARIAS, Charles River Analytics, Cambridge, MA

ANNE S. MAVOR, Study Director
JERRY KIDD, Senior Adviser
SUSAN McCUTCHEN, Senior Project Assistant

COMMITTEE ON HUMAN FACTORS

WILLIAM B. ROUSE (*Chair*), Enterprise Support Systems, Norcross, GA
TERRY CONNOLLY, Department of Management and Policy, College of Business and Public Administration, University of Arizona, Tucson
WILLIAM C. HOWELL, American Psychological Association Science Directorate, Washington, DC
DAVID C. NAGEL, AT&T Laboratories, Basking Ridge, NJ
BENJAMIN SCHNEIDER, Department of Psychology, University of Maryland
LAWRENCE W. STARK, School of Optometry, University of California, Berkeley
EARL L. WIENER, Department of Management Science, University of Miami
GREG L. ZACHARIAS, Charles River Analytics, Cambridge, MA

ANNE MAVOR, Director
JERRY KIDD, Senior Adviser
SUSAN McCUTCHEN, Senior Project Assistant

Acknowledgments

This report is the collective product of the entire panel, and we would like to thank the members for their active contribution in drafting sections of chapters, leading discussions, and/or reading and commenting on successive drafts.

Many individuals have made contributions to the panel's thinking and to various sections of this interim report by serving as presenters, advisers, and liaisons to useful sources of information. First, we would like to thank our sponsors, Judith Dahmann, James Heusmann, and Peter Polk of the Defense Modeling and Simulation Office, for their interest in the topic and their guidance to the panel. In addition, we would like to extend our appreciation to John Laird (University of Michigan), Paul Lehner (MITRE Corporation), H. Kent Pickett (TRADOC Analysis Center, Leavenworth, KS), Douglas Reece (University of Central Florida), and Barry Smith (NASA Ames Research Center, Moffett Field, CA), who provided us with invaluable information concerning current modeling efforts in the military and at NASA.

Finally, staff at the National Research Council made important contributions to our work in many ways. We would like to express our appreciation to Susan McCutchen, the panel's senior project assistant, who was indispensable in organizing meetings, arranging travel, compiling agenda materials, managing the exchange of documentation across the panel and with our project sponsors, and manuscript preparation. We are also indebted to Christine McShane, who edited and significantly improved the report, and to Jerry Kidd, who generously shared his wealth of knowledge and experience.

Richard W. Pew, *Chair*
Anne S. Mavor, *Study Director*

Contents

Representing Human Behavior in Military Simulations

Executive Summary

This interim report reflects the first of a two-phase effort in support of the Defense Modeling and Simulation Office (DMSO) project to review the state of the art in human behavior representation as applied to military simulations. It presents an overview of the needs for models of human behavior, summarizes a methodology for ensuring development of useful models, and describes selected psychological process models that can improve the realism with which human-influenced action is represented.

The need to represent the behavior of individual combatants as well as teams and larger organizations is expanding as a result of increasing use of simulations for training, systems analysis, systems acquisition, and command decision aiding. Both for training and command decision aiding, the behaviors that are important to represent realistically are those that can be observed by the other participants in the simulation, including physical movement and detection and identification of enemy forces. It is important that observable actions be based on realistic decision making and that communications, when they originate with a simulated unit, be interpretable as the result of sensible plans and operations. A team should manifest the range of behaviors required to be consistent with the degree of autonomy it is assigned, including detection and responding to expected and unexpected threats. It should be capable of carrying out actions on the basis of communications typically received from its next higher-echelon commander.

In the panel's view, achieving realism with respect to these observable outcomes requires that the models be based on psychological and sociological theory. For individual combatants, it is important to represent the processes underlying the observable behavior, including sensing and perception, situation awareness, working memory, long-term memory for representing declarative knowledge, decision making, and task management. For larger units, it is important to represent the organizational structure as well as the products of those structures.

Added realism also can be achieved by representing a number of individual and organizational behavioral moderators. Moderators at the individual level, such as workload and emotional stress, serve to enhance or degrade performance, as reflected in the speed and accuracy of performance. Moderators at the organizational level, including the average level of training, whether standard operating procedures are followed, the level and detail of those procedures, and the degree of coupling between procedures, all affect performance.

It is also important, in the panel's view, to introduce learning and intrinsic variability into human behavior representations. Not only will learning allow human behavior representations to respond realistically to changes in the environment without having to anticipate every possible change ahead of time, but it will also allow the models to more closely reflect intelligent human behavior. Furthermore, capturing different levels of learning will allow the representation of different levels of skill, from novice to expert.

Real human behavior is inherently variable and unpredictable, sometimes even irrational. The same person does not always respond identically to identical situations. Although this kind of variability is supported by learning models, it may also be appropriate to introduce controlled randomness into human behavior representations. Trainees who are repeatedly exposed to the same simulations can be expected to learn their idiosyncrasies. The addition of learning and variability will tend to focus them on valid general rules.

When undertaking development of human behavior representation, the panel suggests that the Defense Modeling and Simulation Office encourage developers to make use of a systematic methodology. Such a methodology should include the following steps. First, they should employ interdisciplinary teams consisting of sociologists and cognitive psychologists, computer scientists who are knowledgeable in the contemporary literature and modeling techniques, and specialists in the military doctrine and procedures of the domain to be modeled. Second, they should review alternatives and adopt a general architecture that is most likely to be useful for the dominant demands of the specific situation. Third, they should review available unit-level frameworks and support the development of a comprehensive framework for representing the command, control, and communication structure. Fourth, they should review available documentation and seek to understand in depth the domain and its doctrine, procedures, and constraints. They should prepare formal task analyses that describe the activities and tasks as well as the information requirements and the human skill requirements that must be represented in the models. Also, they should prepare unit-level task analyses that describe such factors as resource allocation and communication protocols. Both individual and unit-level analyses should be guided by the computational structure that will encode the tasks. Fifth, to prepare estimates of the parameters and variables included in the model that are unconstrained by the domain or procedural requirements, they should use results from the behavioral research literature, procedural model analysis, ad hoc experimentation, and, only as a last resort, expert judgment. Finally, they should systematically test, verify, and validate the behavior and performance of the models at each stage of development. Methods for achieving validation of large, complex models are themselves in need of further research exploration.

In the final section of the report the panel identifies several promising substantive areas in which modeling technologies are sufficiently advanced that developers of human behavior representations should investigate and seek to incorporate them in their models. In the area of individual learning, memory, and retrieval, three technologies are discussed—production rules, exemplar, and neural network models; for unit-level learning, the technologies include simulated annealing, neural networks, and genetic programming. In the area of attention and performance, the state of research on low-level attention (governing perception and motor actions) and higher-level planning and decision making are reviewed. In the area of decision making at higher command levels, models derived from game theory, dynamic decision trees, and multiattribute decision theory are presented along with current approaches in operations research, engineering, and organizational science. At the individual combatant level, the discussion focuses on dynamic decision models that allow for individual differences in decision strategy and motivational factors. In the area of situation awareness, there have been fewer attempts to generate computational models, except for some preliminary work on prescriptive models based on belief networks. At the organizational level, most computational models have been multiagent models. They range from symbolic distributed artificial intelligence models to complex adaptive agent models using techniques from genetic algorithms, neural networks, and simulated annealing.

In this first phase of its work, the panel concludes that: (1) human behavior representation is essential to successful use of both war games and distributed interactive simulation; (2) current human behavior representations can be improved by incorporating the knowledge of what is already known from the social and behavioral sciences, cognitive science, and human performance modeling; and (3) great additional progress can be expected through the funding of new research and through the application of existing research in areas that the panel is charged to identify both in this interim report and in its final report.

In the second phase of the project, the panel will conduct a more thorough review and analysis of the theoretical and applied research of human behavior modeling as it applies to the military context at the individual and larger-unit levels. The panel's final report will recommend a research and development agenda to move the field forward in a systematic and integrated manner. The panel recognizes that its initial articulation of needs in this first phase is broad.

Drawing on the military background and knowledge of our sponsor, we plan, as part of the second phase, to conduct a more complete analysis of the strengths and weaknesses of current models and assess the feasibility of establishing priorities for human behavior representation in military modeling and simulation activities.

1

Introduction

In response to a request from the Defense Modeling and Simulation Office, the National Academy of Sciences/National Research Council (NAS/NRC) is undertaking an 18-month study of the state of the art in human behavior representation as it applies to military simulations. Working within the context of the requirements established by military simulations, the panel is reviewing and assessing whether and how well the processes and effects of human behavior at the individual, unit, and command levels can be incorporated into these simulations.

A two-phase effort is under way. This report is the result of the first phase, which has been conducted over an eight-month period. As an introduction to its task, panel members attended a set of three workshops sponsored by the Defense Modeling and Simulation Office, at which leading military contractors described their efforts to model human behavior for a variety of military simulations. The panel has also had extended discussions with military modelers and, taking advantage of the expertise found in its membership, has begun an intensive exploration of the scientific domain of human behavior to identify areas in the literature that are most pertinent to the problems of military modeling. Based as it is on a relatively brief exposure to the problem area, the report provides a preliminary assessment of the needs of military users, suggests an approach to developing human behavior representations, and identifies some promising recent developments. The material presented here should be considered a statement of the panel's progress at the midpoint of its work. We attempt to summarize our present views concerning the relationship between human behavior representation and military simulations and to sketch out at least the beginning of an overall research strategy, to be refined and elaborated in the final report.

In a second phase, the panel will conduct a more thorough review and analysis of the pressing issues in the current generation of battle management simulations and the theoretical and applied research of human behavior modeling as it applies to the military context at the individual and unit levels. The report prepared at the end of the second phase will recommend a research and development agenda designed to move the field forward in a systematic and integrated manner.

This final report will offer guidance to researchers and practitioners who are developing military simulations as well as to those individuals who are responsible for providing the research and development framework for future military simulation activities.

Discussion of human behavioral representation among the experts working in this domain ranges much more broadly than is represented by the charge to this panel. The panel's focus is on the technology and knowledge needed to develop useful and usable human behavior models, ranging from the individual combatant to battalion and brigade-level units. The panel also discusses the analysis required as a prerequisite for model development, as well as verification and validation, because they are important to the generation and success of such models. The state of the art of computer technology, scenario generation mechanisms, and human interfaces to the models themselves are considered outside the panel's scope and are not discussed.

On the basis of its work so far, the panel concludes that: (1) human behavior representation is essential to successful use of both war games and distributed interactive simulation; (2) current human behavior representations can be improved by incorporating knowledge of what is already known from the social and behavioral sciences, cognitive science, and human performance modeling; and (3) great additional progress can be expected through the funding of new research and through the application of existing research in areas that the panel is charged to identify in this interim report and in its final report.

WHAT IS HUMAN BEHAVIOR REPRESENTATION?

In general, the term *model* has different meanings in different communities. For some, a model is a physical replica or mock-up. For others, a model can be a verbal or analytical description or a block diagram with verbal labels. The term *human behavior representation* has been coined by the defense modeling and simulation community to refer to models of the human behavior or performance that needs to be represented in military simulations. In today's armed services, human behavior representations are being used for many different purposes. The prime beneficiaries of improved behavioral representations are the users of simulation, who can be categorized according to the specific application of the simulations they use. *Training simulation users* are instructors and trainees who use simulations for instruction. *Mission rehearsal simulation users* are operational forces who use simulations to prepare for specific missions. *Analysis simulation users* employ their simulations to evaluate alternative weapon systems, doctrine, and tactics. *Acquisition simulation users* are those who must make acquisition decisions based on the anticipated performance of weapons systems. *Command decision-aiding simulation users* are commanders who wish to evaluate alternative courses of action before committing forces. Of course, the modelers who implement the models also stand to benefit from the availability of improved representations.

Two primary critics will view the outputs of human behavior representation and judge whether they are credible or not. First, participants in real-time battlefield simulations will see the performance of individual soldiers and higher-level units in terms of the individual and unit behavior they exhibit, the execution of plans they formulate, and the battle outcomes that result. Second, players in non-real-time battlefield war games will observe only the resulting movements of troops and units, attrition results, and battle outcomes. The details of how individual humans actually accomplish tasks—that is, the psychological theory—is at this level irrelevant as long as the behavioral outcomes seem realistic. Similarly, many details about how groups accomplish tasks, such as group-level decision-making details, may be irrelevant. However, other details—such as, how communication works and the role of the command structure—may be important if the behavioral outcome is to be accurate. What is meant by realistic is open to interpretation and is discussed in more detail below.

In this report we use the term *human behavior representation* to mean a computer-based model that mimics the behavior of either a single human or the collective action of a team of humans. It may be in the context of a self-contained computer simulation used to simulate a battle and run once or many times to produce outputs that reflect the battle outcomes, either individually or statistically. Or it may be a real-time simulation of the behavior of selected battlefield elements that can be viewed by real crews performing in other battlefield element simulators, such as squads of individual soldiers, ground vehicles, and aircraft, to allow the battle to be played out in the simulated world interactively.

To review the state of the art in human performance modeling, with specific attention to potential military applications under the purview of the Defense Modeling and Simulation Office, is especially challenging because the ways in which the models will be used are substantially different from the goals and purposes of typical academic researchers studying and modeling human performance. Most academic researchers concerned with human performance are interested in the interplay between empirical data (experimental, field, or archival) and theory. They implement their theories in terms of executable models so that the detailed assumptions will be revealed and so that they can validate and evaluate the implications of the theory. Their theories are at the individual level and are typically about specific human performance capacities and limitations, such as attention, decision making, and perceptual-motor performance. Rarely do these researchers articulate a comprehensive model of human performance that will behave in the aggregate like real humans. Nevertheless, that is the challenge presented by the requirements of military simulations.

At the unit level, theories typically involve group performance and the way in which it is affected by the communication among group members, procedures, command and control structures, norms, and rules. Many of these theories can be articulated as computational models, which often illustrate the type of impact that an isolated change in procedures or structures may have. These models may generate realistic unit-level behavior; they are rarely accurate, however, at the individual level.

The panel has been and will continue to be challenged by the need to focus on behavioral outcomes and to connect knowledge and theory of human behavior with realistic behavioral outcomes rather than becoming bogged down in details of theory. However, it is the panel's underlying concern that achieving the desired outcomes with both realism and generality requires the models to be based on the best psychological and sociological theory available. In fact, this is a limitation of the current modeling efforts the panel has reviewed. In the absence of such theory, the models are brittle in the sense that mild deviations from the conditions under which they were created produce unrealistic behavior and simplistic responses that do not correspond to the behavior of real individual soldiers or units. To avoid this brittleness and lack of correspondence between the model and real behavior, it is necessary to approximate the underlying psychological structure correctly. For example, in a simulation of the behavior of a flight of attacking helicopters, the helicopters were moving out to attack. One was designated the scout and moved ahead out of sight while the others hovered, waiting for a report. The scout was shot down. Having no further instructions, the others continued hovering until they ran out of fuel. The model could obviously be fixed to eliminate this specific bug in the program, by concatinating further if-then rules. However, what is really needed is a more general decision-making process for the lead pilot that can select among alternative courses of action when expected information does not become available as needed.

Existing theory is the best means we have to define such situations. We also discuss models of learning as frameworks within which to build specific behavioral models. These models are often based not on an explicit theory, but on an architecture that is broader than the specific behavior to be simulated. However, operating within these frameworks may be helpful and important in minimizing brittleness. We discuss probabilistic and stochastic modeling frameworks that can help capture the unpredictability and variability of real behavior.

SETTING EXPECTATIONS IN THE USER COMMUNITY

In discussions the panel has had with various representatives of the user community, it has become clear that there is wide variation in the expectations of what users believe is possible in terms of generating human behavior that is realistic, creative, adaptive, and consistent with military doctrine. We strongly suspect that what can be achieved in the near term is much more limited than what some potential users expect will be possible. One purpose of this project is to elaborate those aspects of performance that the panel determines are implementable now, those aspects that require significant translation of scientific theory and principles to develop into components of computer-based behavioral models, and those aspects of behavior for which the behavioral and social sciences provide inadequate knowledge for developing realistic models in the near future. The goal is to describe in general terms the theoretical and operating principles of models that are applicable to human behavior representation over varying time horizons, from immediately to three to five years; to describe the research necessary to translate current theory into future models; and to describe the areas in which models are needed, but for which theory must be further developed before the users can expect to have models that adequately meet their needs. These objectives will be addressed in the final report.

A fundamental problem that faces the human behavior representation community is the assessment of which of the many modeling requirements will make a difference in the resultant quality of the models, depending on the intended use of the simulation. As the panel has deliberated, it has become clear that a panel of social and behavioral scientists, operating as a consultative group, cannot set these priorities without much more experience in dealing with the specifics of concern to the military community. We may be able to say what requirements will produce models that will more accurately characterize the somewhat abstract human processes of attention, memory, or decision making, but that is different from specifying the requirements that will produce models that appear to simulation users to be more like real individual combatants and military units. It is just not known which of the many improvements in human behavior representation will really make a difference in how a modeled combatant will be viewed, in terms such as meeting the expectations of the opposing force or minimizing the ability to "game" or learn the internal workings of the simulation. This is an aspect of the classic problem of simulation fidelity. Analysts would like to have high fidelity only where it matters, but for representing humans in combat simulations, no one has shown clearly just what aspects of fidelity matter.

2

Modeling Needs for Human Behavior Representation

WHAT ARE THE MODELING NEEDS?

The Under Secretary of Defense for Acquisition and Technology sets as an objective to "develop authoritative representations of individual human behavior" and to "develop authoritative representations of the behavior of groups and organizations" (U.S. Department of Defense, 1995:4-19–4-21).

Presentations made at three workshops held by the Defense Modeling and Simulation Office, formal briefings to the panel, and informal conversations among panelists, Department of Defense representatives, and Department of Defense contractor personnel suggest that users of human behavior representation do not consider the current generation of these representations to reflect the scope or realism that is required for the range of applications of interest. The panel interprets this as a need for representation of larger units and organizations, as well as for better agreement between the behavior of modeled forces (individual combatants and teams) and that of real forces; for less predictability of modeled forces, to prevent trainees from gaming the training simulations; for more variability due not just to randomness but also to reasoned behavior in a complex environment and for realistic individual differences among human agents; for more intelligence to reflect the behavior of capable, trained forces; and for more adaptivity to reflect the dynamic nature of the simulated environment and intelligent forces.

Levels of Aggregation

Authoritative behavioral representations are needed at different levels of aggregation for different purposes. At various times, representations are needed for:

- Individual combatants, including dismounted infantry,
- Squad, platoon, and/or company,
- Individual combat vehicles,
- Groups of combat vehicles and other combat support and combat service support,
- Aircraft,
- Aircraft formations,
- The output of command and control elements, and
- Large units, such as Army battalions, brigades, or divisions, Air Force squadrons and wings, and Navy battle groups.

They are needed for OPFOR (opposing forces or hostiles), BLUFOR (own forces or friendlies) to represent adjacent units, and GRAYFOR (neutrals or civilians) to represent operations other than war and for the interactions among these forces.

The panel recognizes that this initial articulation of needs is very broad, diffuse, and unranked. More structure and prioritization will be needed to drive the development of the program plan to be presented in our final report. However, the panel as it is constituted lacks the military background and knowledge required to establish broad-based priorities for military human behavior representation requirements. This will be addressed by the panel in collaboration with the military sponsors in the final report.

Observable Behaviors

When viewed from the perspective of the simulation user (exclusive of developers) the characteristics of behavior that are visible and interpretable to the users of a simulation depend on the level of aggregation at which the behavior is defined. We consider first the actual user who views a simulation as an individual player, either dismounted or associated with a vehicle. It may be the individual combatant, a ground vehicle or air system commander, a squad or platoon leader, or a commander at a higher level. This individual may be observing a unit at different levels of aggregation as well. Issues of unit-level modeling, aggregation, and scalability are discussed later in this chapter.

The most obvious behavior to be observed is the *physical movement* in the battlespace. It must be at an appropriate speed, and the path followed must make sense in light of the current situation and mission.

The *detection and identification* of enemy or friendly individual units by the human behavior representation must appear reasonable to the observer. The visual search should depend on awareness of the situation, current task demands, and external environmental factors such as the field of view, distance, the weather, visibility, the time of day, the display mode (unaided vision versus night vision goggles), etc.

Decision-making outcomes should reflect situation awareness and real environmental conditions. The decisions concern such observations as which way to move given the plan and the situation presented by the opposing forces; they also concern whether to shoot, seek cover (evade in the case of aircraft or ship), or retreat. Movement decisions should be consistent and coordinated with the behavior of others in the same unit. Decisions should be consistent with the currently active goals. Ideally, individuals would exhibit behavior that reflects rational analysis and evaluation of alternative courses of action, including evaluation of alternative enemy actions given the context. In practice, in time-critical, high-stakes situations, individual decisions are more likely to be "recognition-primed," that is, made on the basis of previously successful actions in similar situations. For example, Klein et al. (1986) showed how experienced fire team commanders used their expertise to characterize a situation and generate a "workable" course of action *without* explicitly generating multiple options for comparative evaluation and selection. More recently, Kaempf et al. (1996) described how naval air defense officers spent most of their time deciding on the nature of the situation; when decisions had to be made about course of action plans, fewer than 1 in 20 decisions focused on option evaluation.

Representation of *communication processes* also depends on the specific purposes of the simulation but should follow doctrine associated with the particular element. Communication need be represented only when it is providing relevant objective status, situation assessment, or unit status information that will affect action at the level of the unit being represented. Communication may take several forms and employ several modes, including direct verbal communication, hand gestures, radio communication, and data link. High-resolution models of small teams may require explicit representation of message content, form, and mode.

HOW PROCESS MODELS OF THE INDIVIDUAL SUPPORT THESE NEEDS

Requirements for behavioral representation at the individual level are accomplished by means of a number of specific components of human performance that include active processes and supporting memories. Realistic models of behavior at the individual level require the representation of at least these components to support the models that execute the behavior. There is room for disagreement among experts concerning the scope and specification of a comprehensive set of components such as these. We claim only that the processes described here are critical determiners of the observable behaviors that are important to military simulations. The relevant components of human behavior are briefly discussed in the following paragraphs.

Sensing and perception are processes that transform stimulus energy into internal representations that can be operated on by cognitive processes. Key elements of individual combatant behavior in military simulations are the detection, identification, and classification of targets, the determination of battlefield situation awareness by observation, and the recognition of patterns of OPFOR movement and positioning. In addition, we consider communications (verbal and gestural) as a special case of sensory input critical to modeling the interactions of individual combatants with team members.

To be consistent with the behavior of real humans, individual combatant models must also exhibit selective attention, a process whereby information processing is focused on a relatively small subset of the sensory stimuli available at any time. The models should account for both visual and auditory attention.

Working memory is the functional component that temporarily holds information for cognitive processing; it is limited in information capacity and persistence. Working memory is significant to military simulation because it is a cognitive bottleneck and therefore limits human performance. For example, a fighter pilot may momentarily be overloaded with competing tasks, forget about the presence of a threat for an instant, and perform an inappropriate maneuver. To the extent that such errors are desirable in modeled behavior, some explicit representation of working memory is essential in an individual combatant model.

Long-term memory is the functional component responsible for holding large amounts of information for long periods of time. There is clearly a need for individual combatant models to represent the storage and retrieval of declarative (factual) knowledge (e.g., weapon characteristics and topography) and procedural knowledge (e.g., tactics and procedures).

Situation assessment is the process by which situation awareness is achieved. It should reflect realistic and available information about the status and position of other friendly forces, own forces, and opposing forces. It should reflect the weather, the mission and objectives of the force, and reasonable inferences based on this information. Ideally it should reflect the expected actions of own and opposing forces in the near future. This implies that the modeled soldier must seek and acquire the relevant knowledge and retain it in memory. Similarly, situation awareness should be changed by observing the results of one's own unit actions. For example, if a particular movement results in unexpected exposure to intense counter fire, that result should revise the assessment of the strength of the opposing force.

Decision making is the process for generating and selecting alternatives and is perhaps the most universally needed process for individual combatant models. Some type of decision process is needed to select the next set of actions based on an evaluation of current information; this process is needed for the selection of actions by individual combatants as well as for high-level command and control.

Task management is the process of managing multiple, concurrent tasks and is ubiquitous in combat operations. It is a cognitive activity that operates at a higher level than attention allocation, since it involves planning and prioritization as well as moment-to-moment task direction. An infantryman confronting an enemy may have to concurrently decide on a general course of action, plan his path of movement, execute tactical movement, and fire his weapon. When engaging multiple targets, a tank crew on an active battlefield must continuously navigate and control the vehicle, search for targets, aim and fire the gun, and assess battle damage. A pilot must simultaneously control the aircraft, plan maneuvers, navigate, communicate with his wingman, control sensors, aim and fire weapons, and monitor and manage other aircraft systems. A commander responsible for several units must continually divide his attention among those units.

Motor response, broadly speaking, refers to the functions performed by the neuromuscular system to carry out the physical actions selected by the above-mentioned processes. Decision making, planning, and other invisible cognitive behaviors are ultimately manifested in observable behaviors that must be simulated, with varying degrees of realism depending on actual applications. Aspects of this realism include response delays, speed-accuracy trade-offs, and anthropometric considerations, such as constraints on movement due to limited reach and strength.

HOW PROCESS MODELS OF THE UNIT SUPPORT THESE NEEDS

The behavioral requirements at the unit level are accomplished by taking into account specific structures and processes. As at the individual level, our claim is simply that these structures and processes are critical determiners of observable unit-level behaviors that are important to military simulations.

The processes and structures that facilitate and limit *communication and coordination* affect many unit-level behaviors, including speed of response and accuracy of response. Furthermore, the vulnerability of the unit when under attack or faced with a novel situation is dependent, in part, on the specific communication and coordination structures and processes. In addition, the likelihood of a unit being observed by OPFOR depends, in part, on the mode of communication chosen and the frequency of communication.

Shared cognition is the set of information and processes held in common by unit members. The extent to which the shared cognition is accurate and is jointly held and the amount and type of information that is jointly held can affect the ability of unit members to coordinate without communication.

Members of the unit act concurrently. The *concurrency* of action can lead to coordination difficulties, particularly when unit members act to achieve their own goals without knowing the goals and constraints of other unit members.

Group decision making, unlike individual decision making, takes into account the role of compromise, negotiation, power, and social influence in determining the joint outcome. Some type of group decision process is needed, even if it is a process that aggregates individual decisions using some weighting function.

Resource allocation is the process of reassigning resources as tasks or goals change. Resource reallocation may be necessitated by depletion of some resources or because some subunits have become incapacitated.

NEED TO REPRESENT BEHAVIORAL MODERATORS

A variety of team- and individual-level factors may have systematic effects on performance and may alter behavior in a number of ways. For example, they may act as multipliers, moderators, or mediators. Or they may interact and influence behavior in complex nonlinear ways.

Excessive workload, fatigue, or generalized emotional stress can be expected to degrade overall performance. The level of achievement motivation will also affect performance. In general, these variables involve individual differences, but performance can also be affected in the aggregate by the quality of leadership and morale. This list is only partial; there is a need to enumerate the full range of the kinds of personality and task moderators that affect overall performance, to operationally define them, and to determine which ones are important for which simulation applications. At the unit level, the factors of average level of training, whether or not standard operating procedures are followed, the level of detail in these procedures, the degree of coupling between procedures, and so forth, can all affect performance.

NEED TO REPRESENT LEARNING

The term *learning* is generally used to mean a change in behavior as a result of experience. In the context of human behavior simulations, it implies that an individual or unit is capable of evaluating the results of its own actions and can change behavior in order to improve the result when the same or a similar circumstance is presented more than once. Current simulation models under review by the Defense Modeling and Simulation Office lack any capability of learning from experience and adapting appropriately to changes in the contingencies in the environment. This causes serious problems for the simulations in many ways. Learning is clearly one of the most important capabilities for survival. An adaptive human behavior representation must be able to learn to anticipate the moves of an opponent in order to ultimately defeat the opponent. An adaptive human behavior representation must also be able to respond appropriately to changes in its model of the environment. Learning is important both within a simulation run and across sequential or related simulation runs.

The addition of learning models to the simulation also provides several other important benefits. First, they offer one method for including individual differences in skill levels. Second, learning provides a principled way to introduce variability into the behavior of the human behavior representation. The action taken by an agent on a new occasion will vary depending on the specific history of consequences from previous actions, and this history will vary in time and across agents.

An effective learning process may also provide support in developing human behavior representations by substituting for some aspects of the knowledge acquisition process. Currently, rules are abstracted from human experts in an extremely costly and time-consuming manner and then are handwritten into the programs. This is a very inefficient and costly way to acquire knowledge. A more efficient and natural way would be to have the simulated agent learn these rules either from examples or by imitation of an expert's behavior. Although we are a long way from building learning models that could replace knowledge acquisition in this way, it is a direction to be explored.

A key concern to be addressed in dealing with human behavior representations that learn is level-of-expertise control and certification. That is, the user will need to know the experience level of the human behavior representation, in terms of what schoolhouse training it has been given, what engagements it has fought, and so forth, to be able to assess battle outcomes in context.

NEED TO REPRESENT VARIABILITY IN DECISION MAKING

The great majority of models that have been used to represent human behavior and decision making in military simulations have employed a fixed and predictable set of rules to govern the next action or decision, for a fixed set of conditions that hold at that time (e.g., MIDAS[1] by the National Aeronautics and Space Administration, SOAR[2]/IFOR by Tambe et al., 1995). Such an approach has many desirable properties: it is relatively easy to understand what the model is doing and why at any given moment; it is relatively easy to catch and repair programming errors; the model is more understandable and transparent; intuitions are easy to develop concerning why the model acts as it does; and when the

[1]MIDAS (Man-machine Integration Design and Analysis System) is a fully constructive operator performance simulation designed to support engineering design decisions.

[2]SOAR (State, Operator, and Result) is a cognitive architecture that attempts to provide a general theory of cognition, not tied to any specific problem space. It provides a set of principles and constraints on cognitive processing, and it is an artificial intelligence programming language.

model needs to be altered, it may be easier to predict what a given change will produce.

Despite these advantages, there are even better reasons to think that effective models will have to incorporate unpredictability in various ways and for various reasons. First, human behavior is inherently variable, unpredictable, and sometimes irrational (March and Simon, 1958; Simon, 1955, 1956; Townsend and Busemeyer, 1995). A model will, therefore, fail to reproduce normal and realistic patterns of actions and decisions unless it incorporates a sufficient degree of random variability—variability that probably needs to be introduced at all levels of the simulation.

Second, the use of simulations for training purposes is harmed by fixed and predictable patterns of simulated behavior. It is easy for a trainee to learn exactly what sequences of actions and behaviors will work in a given class of settings, based not on valid general rules, but on idiosyncratic details that are fixed and immutable properties of a given simulation.

Third, even when human behavior can be modeled accurately at a microanalytic level, the observed behavior at a macroanalytic level may appear to be randomly variable. It can save enormous programming effort, and still capture the essence of behavior, by introducing variability macroscopically. This approach is also useful when the individual micro mechanisms that drive behavior are unknown and/or in debate. Similarly, game theory has identified many situations in which there is no rational strategy (such as an extended-play prisoner's dilemma with a fixed number of games against the same opponent—e.g., Myerson, 1991); behavior in such situations can appear to be and might well be quite probabilistic. Related to such situations are those with infinite planning regressions; for example, one's plans might depend on what one assumes the opponent knows and will do based on his or her knowledge of what one knows and will do, etc. Finally, there are well-known demonstrations in nonlinear systems of chaotic and pseudochaotic behavior of simple fixed rules applied in iterative fashion—as time proceeds, the present state of the system becomes less and less predictable from the starting states of the process (Beltrami, 1987); human behavior in general is certainly nonlinear, and undoubtedly shares this property. In all these cases, the unpredictability of behavior can be captured relatively simply and reasonably accurately by the introduction of random or stochastic noise in the simulation.

Fourth, mathematical models and computer simulations of human behavior have been carried out since the 1950s by cognitive psychologists (and, more recently, cognitive scientists). Great dividends have accrued through the introduction of stochastic variability and probabilistic choice. In the 1980s, it became fairly common to propose fixed and deterministic neural network and connectionist models, and the general complexity of such systems (the large number of connections and nodes and the difficulty of anticipating the macro changes that occur during learning) seemed to provide adequate accounts of variable human behavior, within certain limits. However, more recently it has become clear that even these systems require stochastic variability (e.g., McClelland, 1993).

Fifth, it is sometimes claimed that fixed and deterministic models are adequate even in the face of known variability in the modeled behavior, to the extent that they represent linear and mean approximations of the underlying behavior. Human behavioral mechanisms, however, are anything but simple and linear, and there are numerous demonstrations that qualitatively incorrect predictions are produced by mean approximations to behavioral distributions. Just one of many examples occurs in behavioral models in which decisions are modeled as random walks or diffusion processes. The kinds of qualitative changes in predictions that can occur through the introduction of variability are well illustrated, for example, by the research of Ratcliff and Van Zandt (1995).

Sixth, at the unit level, learning can interact with various characteristics of the unit (such as the form of the command structure) in nonobvious ways. For example, the allocation of resources can affect what skills, perspectives, and problem-solving strategies are learned (Brewer and Kramer, 1985), which can, in turn, affect the unit's flexibility in a crisis (Carley, 1991b). As another example, spatial or command and control structure can enable faster and more robust unit-level learning (Collins, 1992).

Finally, present models and simulations used by the services, for the most part, do not take into account personality variables that are well known to influence behavior in important ways (e.g., cautious and adventurous decision makers may choose very different actions). One approach to this problem involves incorporation of personality variables into the models. This would increase the complexity of the models and would require knowledge of personality factors that may go beyond the present state of knowledge. It may be possible to apply a much simpler approach that could lead the models to produce appropriate variability in behavior simply through the addition of variability, probabilistic choice, and stochastic noise.

These arguments in favor of the incorporation of noise and probabilistic variability in the models appear compelling. Nevertheless, such changes in approach do not come without cost. When these features are incorporated, it is harder to tell why the model produces its behaviors, and error correction during model development is much harder. There will be a similar need for certification of the level of variability represented in a particular human behavior representation. Finally, the running of a noisy and probabilistic simulation requires essentially the same computer time as a fixed and deterministic simulation. This area will be further explored and elaborated in the panel's final report.

NEED FOR TEAM BEHAVIOR REPRESENTATION

A team should manifest the range of behaviors required to be consistent with the degree of autonomy it is assigned, including detection and responding to expected and unexpected threats. It should be capable of carrying out actions on the basis of communications typically received from its next higher-echelon commander.

Much of the research on team or small unit-level adaptation leads to the following conclusions. Social dynamics, both equilibrium and nonequilibrium, depend on the rate of agent learning (adaptation or evolution) (e.g., de Oliveira, 1992; Collins, 1992; Carley, 1991a). Various constraints on agent actions can enable faster and more robust unit-level learning (Collins, 1992; Carley, 1992). Social chaos is reduced by having intelligent adaptive agents determine their actions using strategies based on observations and beliefs about others (Kephart et al., 1992) or by having their actions constrained by their position in the unit. This suggests that cognition and the command and control structure both play a defining role on emergent phenomena (Carley, 1992, forthcoming).

Many current military models and simulations that portray actions at the team or small-unit level are inflexible and cannot adapt to different environments. Further, unit-level models of opposing forces are static and reflect performance and decision outcomes based on the same behavior as BLUE forces. Developing models of different forces takes a considerable amount of time. Unit-level learning models have the potential to change this. They can be used to examine the relative effectiveness of one form of command and control when pitted against opposing forces with different command and control forms. Unit-level learning models may also be particularly useful in examining issues of unit effectiveness and flexibility in a dynamic and volatile environment.

When one is modeling teams or units, it is important to consider the following research findings. For example, it has been repeatedly demonstrated that there is no single correct command and control structure for all tasks (Lawrence and Lorsch, 1967); rather, specific structures vary in their effectiveness depending on the environment and training of the humans in the unit (Carley and Lin, forthcoming a). How the unit should be organized depends on the specific tasks being done, the volatility of the environment, the extent to which humans move into and out of various jobs in the unit, the amount, quality, and accuracy of information available for making decisions, and the level and type of training or experience acquired by the participants (Malone, 1987; Roberts, 1990). Command and control structures tend to change over time whether or not the environment is changing (Cohen et al., 1972). These changes often mean that lessons learned in previous engagements are no longer useful. Finally, the command and control structure is an amalgamation of structures, so no single measure adequately captures its behavior.

NEED FOR LARGE-UNIT BEHAVIOR REPRESENTATION

The representation of larger units, such as squadrons and battalions, brigades and wings, and divisions and higher echelons, is manifest in the structures through which information is disseminated and communications that are transmitted to other relevant units, either higher or lower. Such communications can include:

- Mission statement,
- Commander's guidance,
- Task organizations,
- Force packages and availability (what forces may be assigned in a particular battle plan),
- Target lists and priorities,
- Air Tasking Orders (detailed assignment of aircraft missions),
- Operational Orders (detailed assignment of ground missions),
- Alternative Courses of Action (candidate COAs to be evaluated),
- Logistics status,
- Transportation schedules and movement tables,
- Intelligence updates and situation reports,
- Weather briefings, and
- Battle damage assessments (results from battle actions).

Models are needed that reflect the impact of alternative communication structures with realistic time lags and accurately represent transmission error probabilities. Similarly, models are needed that reflect alternative command and control structures. Current military simulations rarely model the command, control, and communication structure.

There are two common approaches to representing the command, control, and communication structure in models that examine unit behavior and decision making. The most common approach is to represent it by using one or more networks (or matrices) representing the linkages among the humans in the unit. For example, a communication structure might be represented as a matrix such that each cell contains a 1 if the row person can communicate with the column person and a 0 otherwise.

This approach could be expanded to the full command, control, and communication structure by specifying the set of matrices needed to represent it. This approach has many desirable properties. First, the representation facilitates measuring many factors that influence unit behavior, such as throughput, breaks in communication, structural redundancy, and workload assignment errors (e.g., Krackhardt, 1994). Second, having commanders or team members specify the structure often makes specific problems in it obvious to them.

A second approach is to use a rule-based representation for procedures. For example, communication protocols could be represented as a series of rules for when to communicate what to whom and how to structure the communication. This approach is complementary to the network approach. In many applications, both the structure and the procedures need to be represented. Representing procedures as rules facilitates linking up unit-level models with models of individual humans, because the unit-level rules can be added to the individual-level models as additional constraints on behavior. For example, this is the approach taken in the team SOAR (Tambe, 1996) and AAIS (Masuch and LaPotin, 1989).

There is a need to represent changes in unit behavior as a function of scalability—this issue is often overlooked. That is, a common assumption made about unit behavior is that the behavior, problems, and solutions differ only in scale, not in kind, as the size of the unit grows. By this line of reasoning, large units of hundreds of individuals should make decisions in the same way and face the same types of problems that small units of 3 to 10 individuals do. Carried to its extreme, this suggests that units and individuals act in the same way, and a unit is just an aggregate of individual-level actions and reactions. This assumption seems unwarranted in many cases. Numerous studies indicate the presence of unit or team effects not accounted for by simple aggregation of individual behavior. Even in a less extreme form, the scalability assumption may not hold in all cases. For example, research in crisis management suggests that disasters are not scalable (Carley and Harrald, 1997).

When scalability can be assumed, then the same model of unit behavior can be used regardless of unit size. When scalability cannot be assumed, then different models are needed for different size units. To date there has been little research on what types of problems, processes, and behaviors are scalable with unit size and what types are not. Determining the factors that are scalable is an important step in determining when new models are or are not needed.

3

A Methodology for Creating Human Behavior Representations

THE ROLE OF PSYCHOLOGY

Scientific psychology has more than a century's accumulation of data, theory, and experience in research concerning basic human abilities. Many of these results are so useful in practical domains that they have disappeared from psychology and become integrated into technology. For example, the design of high-fidelity audio equipment is based on precise measurements of human auditory abilities collected many years ago (Lindsey and Norman, 1977). Similarly, a variety of other practical domains have been making use of other aspects of psychological research. The development of practical human behavior representations for military simulations is especially exciting because it presents an opportunity to construct and apply comprehensive models of human abilities that span the various subareas of psychology. The resulting synthesis of results and theory will be not only practically useful, but also a stimulus to a broader and deeper theoretical integration that is long overdue.

In addition to a century of research, psychology also has about three decades of experience with computational theories of human ability. Prior to this time, most psychological theory was expressed as verbal descriptions of mental processes whose implications were hard to define because of the lack of precision. The rise of *information-processing theory* in psychology after World War II was helped considerably by applying a metaphor: humans process information in a manner analogous to computer systems. Information is acquired, manipulated, stored, retrieved, and acted on in the furtherance of a given task by distinct mechanisms. The metaphor was taken further in the 1970s by expressing theories and models of mental processes in terms of computer programs. By writing and running the programs, researchers could explore the actual implications of a theoretical idea and generate precise quantitative predictions from the theory.

THE ROLE OF SOCIOLOGY AND ORGANIZATIONAL SCIENCE

Similarly, sociology and organizational science have almost a century of data, theory, and experience in research concerning the behavior of groups of humans. Many of these results are also useful in practical domains and so have disappeared from these fields and become integrated into operations research techniques and best management practices. For example, shop floor allocation procedures derived from early work on scientific management. The development of practical applications of human behavior representations is exciting because it presents an opportunity to construct and apply comprehensive models of units that span distributed artificial intelligence, organizational science, sociology, small group psychology, and political science studies of power.

Following from early work in cybernetics, an information-processing tradition arose within sociology and organizational science. This movement arose more or less in parallel with the one in psychology. However, unlike the movement in psychology, which concentrated on how the individual human acquired, manipulated, stored, retrieved, and acted on information, the movement in sociology and organizational science concentrated on how cognitive, temporal, physical, and social constraints limited the acquisition of information and the consequent actions taken by individuals and groups. The part of this tradition that focused on temporal, physical, and social constraints became known as structural theory. Sociologists and organizational theorists further found that who you know and interact with, and what you think others think of you, affect your opinions, attitudes, and actions. Social information-processing theory, and the various mathematical models of exchange and influence, developed from this research.

Many social and organizational theories are expressed as verbal descriptions of institutional, social, and political processes. As with such descriptions of psychological theories the implications of these processes are hard to determine particularly for dynamic behavior. The reason it is difficult to derive a consistent set of predictions from these verbal models for dynamic behavior is in large part because the behavior of units is extremely nonlinear, involves multiple types of feedback, and requires the concurrent interaction of many adaptive agents. Humans, unassisted by the computer, are simply not good at thinking through the implications of such complexity.

In addition to almost a century of research, sociology and organizational science have about four decades of experience with computational theories of unit-level behavior. Most of these computational models developed from work in information processing, social information processing, and structural theory.

By writing and running these computational programs, researchers can explore the actual implications of theoretical ideas and generate quantitative predictions for unit-level behavior. Also, such models can be used to examine the impact of alterations in group size on the resultant outcome.

THE NECESSITY OF SITUATION-SPECIFIC MODELING

The panel notes that we are a long way at the present time from having a general-purpose cognitive model that can be incorporated directly in any simulation and prove useful. However, the field has developed far enough that simulations incorporating known models and results of cognition and behavior will greatly improve present efforts by the military, if—and only if—the models are developed and exactly tailored to the demands of a given task and situation.

At the present state of development of the field, it is probably most useful to view a human operator as the controller of a large number of programmable components, such as sensory, perceptual, motor, memory, and decision processes. The key is the idea that these components are highly adaptable and may be tuned to interact properly to handle the demands of each specific task in a specific environment and situation. Thus, we may think of the system as a framework or architecture within which numerous choices and specializations must be made when a given application is required. A number of such architectures have been developed and provide examples of how one might proceed, although the field is still in its infancy and it is too early to recommend a commitment to any one architectural framework. Examples include the SOAR architecture (Laird et al., 1986, 1987; Newell, 1990), the ACT[1] series (Anderson, 1976, 1983, 1993), and EPIC[2] (Meyer and Kieras, in press; Kieras and Meyer, in press).

The panel cannot overemphasize how critical it is to develop situation-specific models within whatever general architecture is adopted. The situations and tasks faced by humans in military domains are highly complex and very specific. Any effective model of human cognition and behavior must be tailored to the demands of that situation. In effect, the tailoring of the model substitutes for the history of training and knowledge by the individual (or unit), a history that incorporates personal training and military doctrine. Some may find it disappointing that a general-purpose human model cannot be produced that can be utilized in a variety of quite different situations, but no such model is likely to be available in the near future.

At the unit level, several computational frameworks for representing teams

[1]ACT (Adaptive Control of Thought) is a theory of learning and problem solving used to build intelligent tutors.

[2]EPIC (Executive-Process/Interactive Control) is a cognitive architecture for computational modeling of human performance. It represents a synthesis of results on human perceptual/motor performance, cognitive modeling techniques, and task analysis methodology.

or groups are emerging. These frameworks at worst supply a few primitives for constructing or breaking apart groups and aggregating behavior. At best, they facilitate the representation of formal structure, such as the hierarchy or team, the resource allocation structure, the communication structure, and unit-level procedures that will be inherited by all team members. These frameworks supply only a general language for constructing a model of how human groups perform tasks and the coordination and communication necessary for pursuing those tasks. Representing actual units requires filling in these frameworks with details for a specific team, group, or unit and for a specific task.

A METHODOLOGY FOR DEVELOPING HUMAN BEHAVIOR REPRESENTATIONS

To develop human behavior representations, the panel suggests that the Defense Modeling and Simulation Office encourage developers to employ a systematic methodology, which should include the following steps. First, they should employ interdisciplinary teams. Second, they should review alternatives and adopt a general architecture that is most likely to be useful for the dominant demands of the specific situation. Third, they should review available unit-level frameworks and support the development of a comprehensive framework for representing the command, control, and communication structure. (The cognitive framework adopted should dictate the way in which command, control, and communication procedures are represented.) Fourth, they should review available documentation and seek to understand the domain, its doctrine, procedures, and constraints in depth. They should prepare formal task analyses that describe the activities and tasks as well as the information requirements and the human skill requirements that must be represented in the models. They should prepare unit-level task analyses that describe resource allocation, communication protocols, skills, and so forth, for each subunit. Fifth, to prepare estimates of the parameters and variables included in the model that are unconstrained by the domain or procedural requirements, they should use behavioral research results from the literature, procedural model analysis, ad hoc experimentation, social network analysis, unit-level task analysis, field research, and, as a last resort, expert judgment. Finally, they should systematically test, verify, and validate the behavior and performance of the models at each stage of development. We also encourage government representatives of the military to work with researchers to define the incremental increase in model performance as a function of the effort required to produce that performance. In the sections that follow we elaborate the four most important of these methodological recommendations.

Employ Interdisciplinary Teams

For models of the individual combatant the development teams should include cognitive psychologists and computer scientists who are knowledgeable about the contemporary literature and modeling techniques. They should also include specialists in the military doctrine and procedures of the domain to be modeled. For team-, battalion-, and force-level models, as well as for models of command and control, teams composed of sociologists, organizational scientists, social psychologists, computer scientists, and military scientists are needed. In this way, the resultant simulations will make effective use of the relevant knowledge and many (partial) solutions that have been developed in cognitive psychology, artificial intelligence, and human factors for analyzing and representing human behavior in a computational format. Similarly, employing sociology, organizational science, and distributed artificial intelligence will ensure that the relevant knowledge and solutions for analyzing and representing unit-level behavior will be employed.

Understand the Domain in Depth and Document the Required Activities and Tasks

The first and most critical information required to construct a model of human behavior for military simulations is information about the task to be performed by the simulated and real humans, in terms of the procedures, strategies, decision rules, and command and control structure required for performing the task. This information does not exist in the science of psychology; it is knowledge that exists only in the military domain. For example, under what conditions does a combat air patrol pilot engage an approaching enemy? What tactics are followed? How is a tank platoon deployed into defensive positions? As in the SOAR/IFOR work, military experts have to supply information about the desired skilled behavior that the model is to produce. The form in which this information is collected should be guided by the computational structure that will encode the tasks.

The first source of such information is military doctrine. Doctrine is "fundamental principles by which military forces guide their actions in support of national objectives" (U.S. Department of the Army, 1993). Behavioral representations need to take account of doctrine (U.S. doctrine for BLUFOR, non-U.S. doctrine for OPFOR). On one hand, doctrinal consistency is important. On the other hand, real forces deviate from doctrine when there is good reason to do so. They may deviate because of a lack of training or knowledge of the doctrine. Or they may deviate on their own initiative, to violate an enemy's expectations. Since doctrine is defined at a relatively high level, there is much room for variability in behavior even while maintaining consistency with it. The degree of doctrinal conformity that is appropriate and the way it is manifest for a given model will depend on the goals of the simulation.

Conformity to doctrine is a good place to start in developing a human behavior representation, because it is written down and agreed to by organizational management. However, reliance on doctrine is not enough. First, it does not provide the task-level detail required to create a human behavior representation. Second, just as there are both official organizational charts and informal units, there is both doctrine and the ways the jobs really get done. There is no substitute for detailed observation and task analysis of real forces conducting real exercises.

The Army has a large-scale project to develop computer-generated representations of tactical combat behavior, such as moving, shooting, and communicating. These representations are called Combat Instruction Sets (CIS). According to the developers (IBM/Army Integrated Development Team, 1993), each CIS should be:

- described in terms of a detailed syntax and structure layout,
- explicit in its reflection of U.S. and OPFOR tactical doctrines,
- explicit in how the CIS will interface with the Semi-Automated Forces simulation software, and
- traceable to doctrine.

Information input for Army CIS development comes from written doctrine and from subject-matter experts at the various U.S. Army Training and Doctrine Command schools who develop the performance conditions and standards for mission training plans. The effort includes battalion, company, platoon, squad, and platform/system-level behavior. At the higher levels, the Mission, Enemy, Troops, Terrain, and Time (METT-T) evaluation process is used to guide the decision-making process. The CISs, like the doctrine itself, should provide another useful input to the task definition process.

At the individual level, although the required information is not part of psychology, the process for obtaining and representing the information is. This process, called task analysis and knowledge engineering, is difficult and labor-intensive, but it is well developed and can be performed routinely by well-trained developers and analysts.

Similarly, at the unit level, although the required information is not part of sociology or organizational science, the process for obtaining and representing the information is. This process includes: unit-level task analysis, social network analysis, process analysis, and content analysis. The procedures are difficult and labor-intensive, often involving field research or survey collection, but they can be performed routinely by well-trained researchers.

At the individual level, task analysis has traditionally been applied to identify and elaborate the tasks that must be performed by users when they interact with systems. Kirwan and Ainsworth (1992:1) define task analysis as:

> A methodology which is supported by a number of specific techniques to help the analyst collect information, organize it, and then use it to make judgments or design decisions. The application of task analysis methods provides the user with a blueprint of human involvement in a system, building a detailed picture

> of that system from the human perspective. Such structured information can then be used to ensure that there is compatibility between system goals and human capabilities and organization so that the system goals will be achieved.

This definition of task analysis is conditioned by the purpose of designing systems. In this case, the human factors specialist is addressing the question of how best to design the system to support the tasks of the human operator. Both Kirwan and Ainsworth (1992) and Beevis et al. (1994) provide detailed discussions of a host of methods for performing task analysis as part of the system design process that can be equally well applied to the definition of human behavior representations in military simulations.

If the human's cognitive behavior is being described, cognitive task analysis approaches that rely heavily on sophisticated methods of knowledge acquisition are employed. Many of these are discussed by Essens et al. (1995). Specifically they report on 32 elicitation techniques, most of which rely on interviewing experts, asking experts to make judgments and categorize material, or reviewing and analyzing documents.

Descriptions of the physical and cognitive tasks to be performed by humans in a simulation are important to guiding the realism of the behavior representations. However, developing these descriptions is time-consuming and for the most part must be done manually by highly trained individuals. Although some parts of the task analysis process can be accomplished with computer programs, it does not seem possible that the knowledge acquisition stage will be automated in the near future. Consequently, sponsors will have to establish timing and funding priorities for analyzing the various aspects of human behavior that could add value to engagement simulations.

At the unit or organizational level, task analysis involves specifying the task and the command and control structure in terms of assets, resources, knowledge, access, timing, and so forth. The basic idea is that the task and the command and control structure affect unit-level performance. Task analysis at the unit level does not involve looking at the motor actions that an individual must make or the cognitive processing in which an individual must engage. Rather, it involves laying out the set of tasks that the unit as a whole must do in order to achieve some goal, the order in which those tasks must be accomplished, what resources are needed to accomplish them, and which individuals or subunits have the necessary resources.

A great deal of research in sociology, organizational theory, and management science has been and is being done on how to do a task analysis at the unit level. For tasks, the focus has been on developing and extending project analysis techniques, such as PERT (Program Evaluation and Review Technique) charts and dependency graphs. For the command and control structure, early work focused on general features such as centralization, hierarchy, and span of control. Recently, however, network techniques have been used to measure and distinguish the formal reporting structure from the communication structure. These various approaches have led to a series of survey instruments and analysis tools. There are a variety of unresolved issues, including how to measure differences in the structures and how to represent change.

Instantiate the Models

Specific data are required to fill in the models. Ideally, the model with its parameters specified would already be incorporated into an architectural framework, along with the more general properties of the human information-processing mechanisms. Parameters for selected sensory and motor processes can and should be obtained from the literature. However, many human behavior representations are likely to include high-level decision-making, planning, and information-seeking components. For these components we are still seeking to define suitable underlying structures, and general models at this level will require further research. However, in many cases, the cognitive activities of interest should conform to doctrine or are highly proceduralized. In these cases, detailed task analyses provide data that will permit at least a first-order approximation of the behavior of interest.

Sometimes small-scale analytic studies or field observations can provide detailed data suitable for filling in some aspects of the models, for example, the time to carry out a sequence of actions such as positioning, aiming, and firing a rifle, targeting and launching a missile. Some of these could be readily measured, whereas others could be approximated without new data collection by using approaches based on time and motion study predictions methods from industrial engineering (Antis et al., 1973; Konz, 1995); Fitts' law (Fitts and Posner, 1967); or GOMS[3] (John and Kieras, 1996; Card et al., 1983). These could then be combined with estimates of perceptual and decision-making times to yield reasonable estimates for human reaction times to be incorporated into military simulations.

It is inevitable that some data and parameter requirements cannot be met by the literature or by modeling and analysis and for which it would be too expensive to conduct even an ad hoc study. In those cases, the developer should rely on expert judgment. However, it is the panel's impression from workshops and conference attendance that expert judgment is often viewed as the primary source. We emphasize that it should be the alternative of *last resort.*

Much of the modeling of human cognition that will be necessary for use in human behavior representations—particularly those aspects of cognition involv-

---

[3]GOMS (Goals, Operators, Methods, and Selection Rules) is a relatively simple methodology for making quantitative estimates of the performance times for carrying out well-structured procedural tasks.

ing higher-level planning, information seeking, and decision making—is not yet available and will require new research and development. At the same time, these new efforts can build productively on many recent developments in the field, some of which are discussed in the next chapter.

Verify and Validate the Model

According to the Defense Modeling and Simulation Office, validation is defined as "the process of determining the degree to which a model is an accurate representation of the real world from the perspective of the intended uses of the model" (U.S. Department of Defense, 1996). Different types of models require different levels of validation. The degree of precision needed is guided by the characteristics of the model and its intended use. For example, some large models have too many parameters to test the entire model; in these cases, an intelligent testing strategy is needed. Sensitivity analysis may be used to provide guidance as to how much validity is needed as well as to examine the contributions of particular models and their associated costs. Carley (1996) discusses several types of models, including intellective and emulation models. Emulation models are built to provide specific advice, so they need to include valid representations of everything that is critical. Such models are characterized by a large number of parameters, several modules, and detailed user interfaces. Intellective models are built to show proof of concept or to illustrate the impact of a basic explanatory mechanism. Simpler and smaller than emulation models, they lack detail and should not be used to make specific predictions.

Validation can be accomplished by several methods, including grounding, calibration, verifying, and harmonizing. Note that we are including verification as a subset of validation. Grounding involves establishing the face validity or reasonableness of the model by showing that simplifications do not detract from credibility. Grounding can be enhanced by demonstrating that other researchers have made similar assumptions in their models or by some form of ethnographic analysis. Grounding is appropriate for all models, and it is often the only level of validation needed for intellective models.

Calibration is used to tune a model to fit detailed real data. This is often an interactive process in which the model is altered so that its predictions come to fit the real data. Calibrating a model occurs at two levels: at one level, the model's predictions are compared with real data. At another level, the processes and parameters within the model are compared with data about the processes and parameters that produce the behavior of concern. All of these processes are relevant to the validation of emulation models.

Verification is used to determine the validity of the model's predictions relative to real data. Generally this is accomplished through statistical or graphical comparisons between the model's results and those in the real world—a key requirement is the availability of real data obtained under comparable conditions.

In verification the focus is on results. Furthermore, in the verification process, the model is not altered.

Harmonization is a set of techniques for determining the theoretical adequacy of a verified computational motel. That is, it is an approach used to show that the theoretical assumptions in the model are well grounded in the real world.

It is important to point out that validation is a labor-intensive process that often requires a team of researchers and several years to accomplish. It is recommended that the work be conducted by trained investigators other than those who develop the computational models. In the military context, the most highly validated models are physiological models and a few specific weapons models. Few individual combatant or unit-level models in the military context have been validated beyond the calibration stage—in fact, many have only been grounded. Validation, clearly a critical issue, is necessary if simulations are to be used as the basis for training or policy making.

Large models cannot be validated by simply exhaustively examining the predictions of the model under all parameter settings and contrasting that behavior with experimental data. Basic research is therefore needed on how to design intelligent artificial agents for validating these models. Many of the more complex models can be validated only by looking at the trends that they predict. Additional research on statistical techniques for locating patterns and examining trends needs to be done, and standardized validation techniques that go beyond those currently used need to be developed. This may in part involve developing sample databases against which to validate models at each of the levels. Sensitivity analysis may be used to distinguish between parameters of a model that influence results and those that are indirectly or loosely coupled to outcomes. Finally, it may be useful to set up a review board for ensuring that standardized validation procedures are applied to new models and that new versions of old models are docked against old versions (to ensure that the new versions generate the same correct behavior as the old ones).

4

Promising Developments in Human Behavior Research

Although still in the early stages of its deliberations, the panel has identified a number of areas in which researchers have produced models of human behavior that hold promise for future consideration by the Defense Modeling and Simulation Office. Although following the methodology described in Chapter 3 is important, we emphasize that a critical need for further advancement of the field is to introduce more realistic representations of psychological and organizational processes into the models. The areas of human behavior research presented in this chapter are those with which the panel is most familiar; they should not be considered a complete list. In the second phase, the focus of the panel's work will be to conduct a more in-depth review of these and other areas for the purpose of recommending a plan for research and development. The following sections provide a brief description of current human behavior modeling research efforts in learning and memory, attention and performance, decision making, situation awareness, and organizational structure.

LEARNING AND MEMORY

Learning, one of the oldest topics in psychology, has recently become one of the most intensively studied topics in cognitive science. It is also currently one of the most studied topics in computational organizational theory. As we said earlier, learning can be defined as the gradual accumulation of knowledge through experience, generally as the result of repeated exposure. Memory has been the focus of extensive modeling efforts for the past 30 years. In memory research the concern is usually the storage and retrieval of information presented only once (or only a very few times), whereas in learning research the concern is usually the training of associations across a fairly large number of trials or examples. However, what is learned must be stored in memory, and learning offers one way of thinking about the process of memory encoding. Thus, there is no clear dividing line between these areas of research. Advances in modeling memory and retrieval are mentioned in this section, although learning through experience may be of the most importance for defense simulation purposes.

Learning

Several different types of learning have been defined in the literature. The most relevant for military purposes is learning the best action to take under a particular environmental condition, based on experience with previous consequences and anticipation of future changes in the environment or task. At the individual level, three basic approaches have been developed for constructing formal or computational models of learning: (1) the *rule-based approach* developed in the traditional artificial intelligence (machine learning) literature; (2) the *exemplar approach*, sometimes referred to by computer scientists as *case-based learning*, developed in the psychological literature; and (3) the *neural network approach*, developed in the cognitive science and engineering literature.

At the unit level, two approaches have been taken: (1) the adaptation of the single unit and (2) the evolution of a set of units. Models of single-unit adaptation use either simulated annealing or neural networks; however, genetic programming can also be used. For evolutionary models, genetic algorithms or combined neural network genetic algorithm procedures have been used. With respect to the military context, these learning models can be applied at different levels. One level is to describe the learning process at the individual agent level. Another is to describe the internal adaptation process within the unit at the organizational or team level. Still another is to describe the evolutionary process for groups of organizations.

The rule-based approach employs a production rule framework for representing knowledge. There are at least four basic mechanisms used to learn new rules with this approach: (1) to strengthen rules that lead to successful outcomes, (2) to discriminate by adding conditions before a rule can be applied, (3) to generalize by removing conditions before a rule can be applied, and (4) to use chunking, a procedure for creating new rules that encapsulate the lessons learned through searching for a solution. Good examples of a rule-based learning approach can be found in Anderson (1990) and in Laird et al. (1986). The advantage of these approaches is that they readily fit into the production rule framework that is currently used in most military simulations.

The exemplar-based approach employs a multidimensional vector space representation of experience with previous episodes of situations, actions, and outcomes. Memory is based on storage of previous examples, with some decay over time. Learning occurs by retrieving stored examples that are similar to the present situation and selecting the action that produced the most successful outcome in the past under the current conditions. Good examples of the exemplar approach can be found in Logan (1988) and Nosofsky and Palmari (in press). The exemplar models have proven to be more successful than rule-based models in experimental studies of learning. The exemplar approach also can be implemented in the rule-based systems without too much difficulty.

Neural network models are inspired by elementary principles of neuroscience. Formally, neural networks are dynamic systems that describe the flow of activation from one field of nodes to another. Information from the environment is represented by a distribution of activation across an input field of nodes, and retrieval occurs when the activation in the network settles to an equilibrium output response state. Learning is achieved by updating the weights that connect one field of nodes to another. Learning is formalized as a gradient descent algorithm designed to maximize an objective function. Some advantages of neural networks are their robustness to noisy inputs, their ability to gradually deteriorate with damage (as opposed to the brittle behavior of rule-based systems), and their ability to react intelligently to novel situations (provided that the situation has some similarity to previous experience). Good examples of neural network models for learning are Plaut et al. (1996) and Kruschke (1992).

Memory and Retrieval

Of central importance for simulation purposes are two facts: (1) new information may be encoded incompletely, poorly, or in a form not conducive to retrieval and (2) new information, poorly encoded information, or well-known information may not be retrieved (especially under time pressure).

One method of classifying models of memory and retrieval also involves sorting these into rule-based, exemplar-based, and neural-network-based models. Examples of rule-based approaches include ACT (Anderson, 1993) and the rational model approach (Anderson, 1990). Neural network models include those of Anderson et al. (1977), Chappell and Humphreys (1994), McClelland and Chappell (1994), Grossberg and Stone (1986), Murdock (1982), and Metcalfe-Eich (1982). Exemplar-based models include those of Hintzman (1988), Raaijmakers and Shiffrin (1980, 1981), and Shiffrin and Steyvers (in press). Possibly the best worked out and easiest to use model at present is the SAM model of Raaijmakers and Shiffrin. However, for many simulation purposes, it may be less important to incorporate any single model than to recognize the variability and imprecision of retrieval.

More generally, learning and memory models may be integrated into a common theoretical framework. The basic principles are the same, but the application to the different types of tasks involves additional assumptions and specifications. For example, exemplar-based models may incorporate the same basic principles for categorization and memory recall concerning storage and retrieval of examples, although the use of this retrieved information will differ depending on whether the task requires a categorization response or a exact reproduction of a list of items.

ATTENTION AND PERFORMANCE

It is virtually a truism that the chief limitation on human behavior in most tasks is the capacity of the human processing system. Since attention is the mechanism by which this capacity is allocated, whether one conceives of a single pool of capacity or multiple and relatively separate capacities, it is essential for accurate modeling that attentional processes be represented properly in human behavior representations. Without attentional constraints, the simulated combatant possesses superhuman capacities that would produce grossly unrealistic training and test environments for humans and overly optimistic results for performance.

During the past 10 years, psychological research has shown that the limitations of capacity can be overcome in large part through training, learning, and experience, all of which develop automatic processes that can produce actions and decisions while demanding far less capacity than is true prior to training (for descriptions see Shiffrin and Schneider, 1977; Schneider and Shiffrin, 1977; Shiffrin, 1988; Logan, 1988). The development of automatic processes is not without cost, of course, because the actions and decisions that result may reflect a lack of flexibility and adjustment to new circumstances, and a proper simulation of human behavior for defense purposes must reflect the tension between the need for flexible and appropriate actions and decisions in light of the local situation on one hand, and the need to free up scarce resources by relying on already learned responses on the other. For example, on the battlefield, soldiers are trained to employ direct fire to engage the enemy. However, there are many situations in which indirect fire is more appropriate. Selecting the indirect fire option requires evaluating the situation and ignoring the automatic, learned response. The need to develop automatic responses can be considered a form of learning, as discussed in the previous section. In this section we touch on promising and current models of attentional allocation.

By far the greatest effort in the field has been aimed at developing models for low-level attention, which governs perception and elementary motor actions. Some additional work has gone into developing models of allocation in short-term memory, although these may not in principle be different. A few partial exceptions to the rule that attentional models are aimed at low-level issues and relatively simple tasks are found in the human factors literature and some military applications (e.g., Wickens, 1984; Schneider and Shiffrin, 1977; Fisk and Eggemeier, 1988).

Even in the perceptual and motor attentional approaches, few general simulation models have been proposed, most models being aimed at very specific tasks and issues. For example, a good deal of effort has gone into the question of the allocation of visual attention in time and space. Much research has shown that many performance deficits once thought to lie in limited perceptual and motor capacities instead are due to inevitable decisional limitations imposed by the structure of particular tasks (e.g., Shiffrin and Gardner, 1972; Palmer, 1994; Sperling and Dosher, 1986). In cases in which attentional allocation does play a role in determining performance, considerable effort has gone into distinguishing models in which attention moves continuously in time and space or jumps in quantal fashion from one time-space location to another (e.g., Sperling and Weichselgartner, 1995). Quite recently some attempts have been made to produce more general models of attention and performance that would better generalize across tasks and settings (e.g., Meyer and Kieras, in press). Also relevant to the discussion of attention and performance are the issues involved in the scheduling of mental processes. For example, some mental tasks must be performed in a serial order (one at a time), whereas other tasks allow for parallel processing (several simultaneously). Various combinations and mixtures of serial and parallel processing are also possible. A good example of work in this area is Liu (1996); for a comprehensive review, see Townsend and Ashby (1983).

Attentional models applying to higher-level planning, problem solving, and decision making, still in their infancy, may be of critical importance for military simulations. Some attentional concepts, however, have begun to appear in decision-making models, in the form of weightings on information, differential use of information at different temporal and spatial removes, and differential actions and decisions under different time pressures. These are described in the next section.

DECISION MAKING

The current military simulation models reviewed by the panel make very little use of sophisticated decision-making mechanisms. Decision making in these models is essentially reduced to checking preference values over a list of proposed actions selected for the current situation. One serious problem with this method is that the preferences have to be acquired from experts for every single rule and situation, which is an extremely costly procedure. Furthermore, this method provides no means for evaluating new actions (without consulting with an expert), making it impossible to evaluate innovative plans generated in response to new contingencies. Another problem with the current approach is that actions are picked deterministically for any given situation, so that an opponent player can easily anticipate and take advantage of an anticipated action.

Incorporating more sophisticated decision models, such as those described below, would allow more general procedures for evaluating new actions. Both normative and behavioral approaches can be used to address two types of decision problems: one is called competitive games (decisions involving several intelligent competing agents), and the other is individual decision making (from the point of view of a single individual agent). With respect to the military context, the literature on individual decision making is most useful for individual combatant decisions; at the level of command and control, the literature on organizational learning is most relevant.

Command-Level Models

In principle, command-level decisions should reflect a fair degree of rationality or optimality with respect to the evaluation of plans and the allocation and reallocation of resources and tasks. At the command level, however, the quality of the decisions is a function of a variety of factors, including situation awareness and constraints imposed by the rules of operation. Some command decisions involve response to an opposing force, others involve coordination with supporting forces. No single theoretical tradition is sufficient or has a comprehensive model for all types of command-level decisions. For example, work in operations research and in distributed artificial intelligence has addressed the issue of dynamic resource allocation. Work in game theory has addressed issues of competition against a single enemy. Work in computational organizational theory has addressed issues of coordination and evaluation of changes in command, control, and communication structures.

Elements of command, such as the need to assess OPFOR operations, make strategic decisions, and reassign resources, need to be represented in team- and unit-level models, as well in models of commanders. These models can be used in different ways and so can require different types of models. For example, if a representation of the commander is to be part of a training model for a subordinate, it may be desirable to model the commander as making decisions that, though satisfactory, are not necessarily optimal. For models of the OPFOR commander, it may be desirable to make the commander act in an optimal manner. Depending on how they are to be used, models of high command thus may or may not be based on optimality or learning models.

An *optimality model* is a model in which the simulated agent is trying to locate the optimal solution for a problem. Game theory and operations resource optimizing functions are examples. A *learning model* is one in which the simulated agent is trying to locate a satisfactory or optimal solution and use past experience and knowledge to guide behavior. There are also constrained optimization models that can be adjusted to act like satisficers, such as simulated annealing.

Depending on the goal, there are a number of tools for approaching this problem. A variety of models essentially try to locate the optimal solution given a set of constraints. One of the earliest to attract attention is game theory. More recent approaches, which may be of greater value with regard to command and control, include dynamic programming, simulated annealing, and genetic programming. Incorporating some form of optimality modeling into models of high command may improve their performance.

Game theory is concerned with rational solutions to decisions involving multiple intelligent agents that have conflicting objectives. Models employing the principle of subgame perfect equilibrium (e.g., Myerson, 1991) could be used for simulating command-level decisions because they provide an optimal solution for planning strategies with competing agents. Decision scientists employ decision trees to represent decisions in a complex dynamic environment. Although the number of stages used to represent the past and future and the number of branches extending from each node in a tree can be large, these numbers must be limited by the computational constraints of the decision maker.

The planning horizon and the branching factor are important individual differences among decision makers. Game theoretic approaches tend to be useful for simple trade-offs and situations in which the payoff function is known. However, for many practical problems, a decision tree approach to the entire problem may not be particularly valuable, because it cannot either be constructed in its entirety, or it becomes arbitrarily complex (the curse of dimensionality), or it requires information that is not available, or it assumes that the choices are static and unchanging. In many instances, however, it is possible to use partial trees for only a part of the decision. That is, the trees can be cut off at a short horizon; game theory can then be applied to the truncated tree. Even the use of partial trees may increase the performance of many systems over using heuristic-based approaches.

When decisions, or parts of decisions, cannot be easily represented as a tree, the value of game theory may diminish. For very complex decisions, representing the decision in game theoretic terms may be extremely cumbersome, often requiring the decision to be represented as a set of games and meta games. Furthermore, specifying the decision as a game requires being able to specify the payoff function. This is not always obvious, particularly for decisions for which the payoff is a function of previous actions. Having to specify the payoff or objective function is, in fact, a problem for most optimization, constrained optimization, and learning algorithms.

A commander's job may involve balancing conflicting objectives in high-risk situations. In rational decision theory, the prescribed model for making risky decisions involving multiple objectives is the *multiattribute utility model* (e.g., Clemen, 1996). The states of the tree are represented by multiattribute consequences (i.e., each state is a vector of values, and each coordinate represents the value achieved on a different objective). In this way, optimal strategies are selected that are designed to maximize a multiobjective criteria (e.g., a weighted sum of values). Empirical research has been concerned with determining the form of the multiattribute function. The weighted sum is just an example and not the only form.

Other research suggests that humans do not behave in this way (Axelrod, 1976; Abelson, 1976). That is, they often cannot specify a criterion function, may attend only to a few criteria at a time, may choose the first satisfactory solution and not continue to search for the optimal solution, and so on. Thus, if the objective is to have a representation that is natural (seems like the way humans represent information) or an artificial agent that is descriptive (acts as a human might) then learning, satisficing, and constraint-based approaches would be called for. If the objective is to be normative, or to find the optimal solution, then more straight optimization, rational decision theory, or game theoretic models might be called for. In other words, different modeling tools are needed to model both how the commander makes a decision and how the commander should make a decision.

Individual-Level Models

Representations of low-level combatant decisions should reflect the fundamental abilities and limitations that are empirically observed in human decision-making research. There is a long history of empirical research that has been concerned with the discovery of lists of human biases and simplistic heuristic rules employed by human decision makers (e.g., Kahneman et al., 1982). Unfortunately, this work on heuristics and biases has failed to accumulate and evolve into any coherent formal or computational model of individual decision making. At best, the list of heuristic rules may be used to guide the selection of condition-action rules programmed into current production rule systems of individual decision making.

More recently, some simple heuristic decision rules have been coded as production rules, and then inserted into computer simulation models for decision making (e.g., Payne et al., 1993). In fact, this type of decision model has already been incorporated into some production rule simulation models used by NASA (e.g., MIDAS). A problem with these models is that they are based on overly simplistic assumptions. For example, the information-processing approach used in MIDAS is based on a strictly serial processing assumption, which is known to be inconsistent with a large body of research on human cognition that emphasizes the importance of parallel distributed processing for fast but accurate cognitive performance.

Progress has also been made on the development of dynamic decision models that are highly relevant to the goals of the Defense Modeling and Simulation Office (Busemeyer and Townsend, 1993; Townsend and Busemeyer, 1995; Grossberg and Gutowski, 1987). These models provide a description of decision making under real time constraints; they include effects of time pressure on the speed and accuracy of decision making; they allow for individual differences in impulsiveness and risk aversion; they are driven by emotional and motivational factors; and they can be used to model the effects of stress resulting from fear of impending negative consequences. Another advantage of these dynamic decision models is that they can be readily integrated with learning models (e.g., Busemeyer and Myung, 1992). Thus, adaptive learning models and dynamic decision models can be synthesized into a general model of learning and motivation for real-time adaptive decision making.

SITUATION AWARENESS

Situation awareness is defined as the individual's state of knowledge or mental model of the surrounding situation or environment. It includes an understanding of the dynamics of the situation and the actions that are expected to take place in the future as well as cues for spatial orientation (Endsley, 1995a). A recent report on tactical displays for the individual soldier (National Research Council, 1997) identifies situation awareness as an element critical to successful performance in the combat environment. Other research also shows that situation awareness dominates tactical planning activities (Fallesen et al., 1992; Fallesen, 1993; Deckert et al., 1994). For example, awareness of uncertain assumptions contributes to more flexible planning. Better use of available information results in more elaborate war gaming. Explicit prediction of events is aided by active seeking of evidence to confirm or reject greater awareness of enemy activities.

Maintenance of situation awareness also plays a key role in high-tempo battlefield activities that are more reactive in nature and involve less planning (e.g., the battle drills noted in the training document prepared by IBM, 1993). Several studies have focused on scenarios and conditions under which the decision maker must make dynamic decisions under high levels of uncertainty, high time pressure, and rapid change. These studies span the theoretical-to-applied spectrum and cover many domains. Endsley (1989, 1993, 1995a, b) and Adams et al. (1995) discuss psychological models of situation awareness and the impact of particular system characteristics on operator workload, attention and memory requirements, and the likelihood of errors. Klein and colleagues (1986) have studied a particular type of decision making predicated on the quick extraction of salient cues from a complex environment and a mapping of these cues to a set of procedures.

A variety of situation awareness models have been hypothesized and developed by psychologists and human factors researchers, primarily through empirical studies in the field, but increasingly with computational modeling tools. Because of the critical role of situation awareness in air combat, the U.S. Air Force has taken the lead in studying the measurement and trainability of situation awareness (Caretta et al., 1994). Numerous studies have been conducted to develop situation awareness models and metrics for air combat (Stiffler, 1988; Spick, 1988; Harwood et al., 1988; Endsley, 1989, 1990, 1993, 1995a; Fracker, 1990; Hartman and Secrist, 1991; Zacharias et al., 1992; Klein, 1994).

TABLE 4.1 Characteristics of Situation Awareness (SA) Models

Class	Features	Advantages	Disadvantages
Descriptive	Data driven Qualitative	Reflect actual SA process Capable of handling complex scenarios Empirical Basis	Lack of predictive capability Provide vague or non-extensible conclusions Do not support computational implementation
Prescriptive	Assumption or theory driven Quantitative	Prescribe "optimum" SA process Support computational implementation Support objective SA metric development	High development cost Limitations in applicability Many validity issues

Situation awareness modeling can be roughly categorized into two efforts: descriptive and prescriptive or computational. Table 4.1 summarizes the major features of the two model classes and presents their relative advantages and disadvantages. Most developed situation awareness models are descriptive. Endsley (1995a) presents a descriptive model of situation awareness in a generic dynamic decision-making environment, depicting the relevant factors and underlying mechanisms. The relationship between situation awareness and numerous individual and environmental factors is explored. Among these factors, attention and working memory are considered the critical factors limiting effective situation awareness. Mental model and goal-directed behavior are hypothesized as important mechanisms for overcoming these limits.

Although the descriptive models are capable of identifying the dependent relationships between subsystem modifications and situation awareness enhancements, they do not support a quantitative evaluation of such relationships when no empirical data are available. There currently exists no descriptive model that has been developed into a computational model, for actual emulation of pilot decision-making behavior in real-time simulation studies.

In contrast to the status of descriptive models, few prescriptive models of situation awareness have been proposed or developed. Early attempts use production rules (Baron et al., 1980; Milgram et al., 1984). In these efforts, the situation awareness model was developed as a forward-chaining production rule system. These models lack long-term memory and an internal mental model; thus, they cannot make use of event histories or event cue information.

Recognizing that situation awareness is fundamentally a diagnostic reasoning process, Zacharias and colleagues (1992, 1994, 1996) and Mulgund et al. (1996a, b) used belief networks to develop prescriptive situation awareness models for two widely different domains: counter-air operations and nuclear power plant diagnostic monitoring. Both efforts modeled situation awareness as an integrated inferential diagnostic process, in which situations are considered as hypothesized reasons, events as effects, and sensory (and sensor) data as symptoms (detected effects). Situation awareness starts with the detection of event occurrences. After the events are detected, their likelihood (belief) impacts on the situations are evaluated by backward tracing the situation-event relation (diagnostic reasoning) using Bayesian logic. The updated situation likelihood assessments then drive the projection of future event occurrences by forward inferencing along the situation-event relation (inferential reasoning) to guide the next step of event detection. A more recent application of this modeling approach is described by Mulgund et al. (1996a) in an effort directed at developing a situation-driven display for fighter cockpits.

ORGANIZATIONAL MODELS

The behavior of units has been modeled with some success for the purpose of testing ideas from organizational theories. Achieving greater realization in military simulations will require major adaptations. However, some of the findings from the tests of organizational theory can contribute to efforts in the areas of flexibility, adaptation, and performance monitoring. For example, research in this area has provided insight into the ways in which units change when faced with stress and should be structured to withstand stress (Carley and Lin, 1994, 1995, forthcoming a). Further work in this area may provide insight into factors affecting the success of joint task forces and coordinated responses by multiple forces.

Most of the computational modeling work on unit-level models employs the use of multiagent models. These models range from the more symbolic distributed artificial intelligence models to models using one of the various complex adaptive agent techniques, such as genetic algorithms (Holland et al., 1986; Macy, 1991a, b; Crowston, 1994, in press), neural networks (Karayiannis and Venetsanopoulos, 1993; Kontopoulos, 1993), simulated annealing (Carley and Svoboda, 1996), chunking (Tambe, 1996), and other stochastic learning models (Carley and Lin, forthcoming a, b; Lin and Carley, forthcoming; Glance and Huberman, 1993, 1994). Some of this work rests on, or is related to, mathematical models of distributed teams (Pete et al., 1993, 1994; Tang et al., 1993) and social psychological experimental work on teams (Hollenbeck et al., 1995). Most of these models perform specific stylized tasks, and many of them assume a particular command and control structure.

There are two types of models commonly used at the unit level. The first, intellective models, are small models intended to show proof of concept for some theoretical construct. Such models enable the user to make general predictions about the relative benefit of changes in generic command and control structures.

The second, emulation models, are large models intended to emulate a specific organization in order to locate specific limitations to that unit's structure. Such models enable the user to make specific predictions about a specific organization. Intellective models provide general guidance and an understanding of the basic principles of organizing; emulation models, which are difficult and time-consuming to build, provide policy predictions, but only for a single case.

In these models, the artificial agents, although attempting to act like human agents, are not perfect human analogs (Moses and Tennenholtz, 1990). The basic research issue is how accurate the agents in these models need to be so that a unit composed of many of them acts like a unit of humans. Some research suggests that complete veridicality at the individual level may not be needed for reasonable veridicality at the unit level (Castelfranchi and Werner, 1992; Carley, forthcoming). Moreover, the cognitive capabilities of the individual agents, as well as their level and type of training, interact with the command and control structure and the task the agents are performing to the extent that different types of agent models may be sufficient for modeling unit-level response to different types of tasks (Carley and Newell, 1994; Carley and Prietula, 1994; Carley and Lin, forthcoming b). There have been a few models of unit-level adaptation that are not multiagent models (for a review, see Lant, 1994). These models typically employ autonomous agents acting as the unit or various search procedures. This work has been useful in demonstrating various points about organizational learning (Lant and Mezias, 1990, 1992).

None of the computational unit-level models is at the point of plugging directly into a current Defense Modeling and Simulation Office platform as the model of the unit. In part this is because existing unit models have either too limited a repertoire of command and control structures or too limited a model of task. However, the network-based representation and the various network measures of structure could be incorporated into some current Defense Modeling and Simulation Office simulations.

References

Abelson, R.P.

1976 Script processing in attitude formation and decision-making. In *Cognition and Social Behavior*, J.S. Carroll and J.W. Payne, eds. Hillsdale, NJ: Lawrence Erlbaum Associates.

Adams, M.J., Y.J. Tenney, and R.W. Pew

1995 Situation awareness and the cognitive management of complex systems. *Human Factors* 37(10):85-104.

Anderson, J.R.

1976 *Language, Memory, and Thought*. Hillsdale, NJ: Lawrence Erlbaum.

1983 *The Architecture of Cognition*. Cambridge, MA: Harvard University Press.

1990 *The Adaptive Character of Thought*. Hillsdale, NJ: Erlbaum.

1993 *Rules of the Mind*. Hillsdale, NJ: Lawrence Erlbaum.

Anderson, J.A., J.W. Silverstein, S.R. Ritz, and R.S. Jones

1977 Distinctive features, categorical perception, and probability learning: Some applications of a neural network. *Psychological Review* 84:413-451.

Antis, W., J.M. Honeycutt, and E.N. Koch

1973 *The Basic Motions of MTM*. Naples, FL: The Maynard Foundation.

Axelrod, R.M.

1976 *Structure of Decision: The Cognitive Maps of Political Elites*. Princeton, NJ: Princeton University Press.

Baron, S., G. Zacharias, R. Muralidharan, and R. Lancraft

1980 PROCRU: A Model for Analyzing Flight Crew Procedures in Approach to Landing. In *16th Annual Conference on Manual Control*. Cambridge, MA.

Beevis, D., R. Bost, B Döring, E. Nordø, J-P. Papin, I.H. Schuffel, and D. Streets

1994 Analysis Techniques for Man-Machine System Design. AC/243 (Panel 8)TR/7. North Atlantic Treaty Organization Defence Research Group.

Beltrami, E.
1987 *Mathematics for Dynamic Modeling*. San Diego, CA: Academic Press.

Brewer, M.B., and R.M. Kramer
1985 The psychology of intergroup attitudes and behavior. *Annual Review of Psychology* 36:219-243.

Busemeyer, J.R., and I.J. Myung
1992 An adaptive approach to human decision making: Learning theory, decision theory, and human performance. *Journal of Experimental Psychology* 121: 177-194.

Busemeyer, J.R., and J.T. Townsend
1993 Decision field theory: A dynamic-cognitive approach to decision making in an uncertain environment. *Psychological Review* 100:432-459.

Card, S.K., T.P. Moran, and A. Newell
1983 *The Psychology of Human-Computer Interaction.* Hillsdale, NJ: Lawrence Erlbaum Associates.

Caretta, T.R., D.C. Perry, and M.J. Ree
1994 The ubiquitous three in the prediction of situation awareness: Round up the usual suspects. Pp. 125-137 in *Situational Awareness in Complex Systems*, Gilson, Garland, and Koonce, eds. Daytona Beach, FL: Embry-Riddle Aeronautical University Press.

Carley, K.M.
1991a A theory of group stability. *American Sociological Review* 56(3):331-354.
1991b Designing organizational structures to cope with communication breakdowns. *Industrial Crisis Quarterly* 5(1):19-57.
1992 Organizational learning and personnel turnover. *Organization Science* 3(1): 20-46.
1996 Validating Computational Models. Social and Decision Sciences Working Paper. Pittsburgh, PA: Carnegie Mellon University.
forthcoming A comparison of artificial and human organizations. *Journal of Economic Behavior and Organization.*

Carley, K.M., and J. Harrald
1997 Organizational learning under fire: Theory and practice. *American Behavioral Scientist* 40(3).

Carley, K., and Z. Lin
1994 Relating shifts in C^2 structure to performance. In *Proceedings of the 1994 Symposium on Command and Control Research.*
1995 Organizational designs suited to high performance under stress. *IEEE–Systems, Man and Cybernetics* 25(1):221-230.
forthcoming a Organizational decision making and error in a dynamic task environment. *Journal of Mathematical Sociology.*
forthcoming b A theoretical study of organizational performance under information distortion. *Management Science.*

Carley, K., and A. Newell
1994 The nature of the social agent. *Journal of Mathematical Sociology* 19(4):221-262.

Carley, K.M., and M.J. Prietula
1994 ACTS Theory: Extending the Model of Bounded Rationality. Pp. 55-88 in *Computational Organization Theory*, K.M. Carley and M.J. Prietula, eds. Hillsdale, NJ: Lawrence Erlbaum Associates.

Carley, K.M., and D.M. Svoboda
1996 Modeling organizational adaptation as a simulated annealing process. *Sociological Methods and Research* 25(1):138-168.

Castelfranchi, C., and E. Werner, eds.
1992 Artificial Social Systems: 4th European Workshop on Modeling Autonomous Agents in a Multi-Agent World. MAAMAW 92. S. Martinoal Cimino, Italy, July 29-31.

Chappell, M., and M.S. Humphreys
1994 An auto-associative neural network for sparse representations: Analysis and application to models of recognition and cued recall. *Psychological Review* 101:103-128.

Clemen, R.T.
1996 *Making Hard Decisions: An Introduction to Decision Analysis, Second Edition.* Belmont, CA: Duxbury Press.

Cohen, M.D., J.G. March, and J.P. Olsen
1972 A garbage can model of organization choice. *Administrative Sciences Quarterly* 17(1):1-25.

Collins, R.J.
1992 *Studies in Artificial Evolution.* CSD-920037, Computer Science Department. Los Angeles, CA: University of California.

Crowston, K.
1994 Evolving novel organizational forms. Pp. 19-38 in *Computational Organization Theory*, K.M. Carley and M.J. Prietula, eds. Hillsdale, NJ: Lawrence Erlbaum Associates.
in press An approach to evolving novel organizational forms. Forthcoming in *Computational and Mathematical Organization Theory.*

Deckert, J.C., E.B. Entin, E.E. Entin, J. MacMillan, and D. Serafaty
1994 *Military Command Decisionmaking Expertise* (Report 631). Burlington, MA: Alphatech.

de Oliveira, P.P.B.
1992 Enact: An Artificial-Life World in a Family of Cellular Automata. CSRP 248, School of Cognitive and Computing Sciences Cognitive Science Research Papers. Brighton, East Sussex, England: University of Sussex.

Endsley, M.R.
1989 Pilot situation awareness: The challenges for the training community. In *Interservice/Industry Training Systems Conference.* Ft. Worth, TX.
1990 Predictive utility of an objective measure of situational awareness. In *Proceedings of the Human Factors Society 34th Annual Meeting.* Santa Monica, CA: Human Factors Society.
1993 A survey of situation awareness requirements in air-to-air combat fighters. *International Journal of Aviation Psychology* 3:157-168.
1995a Toward a theory of situation awareness in dynamic systems. *Human Factors* 37(1):32-64.

1995b Measurement of situation awareness in dynamic systems. *Human Factors* 37(1):65-84.

Essens, P., J. Fallesen, C. McCann, J. Cannon-Bowers, and G. Dorfel

1995 COADE—A Framework for Cognitive Analysis, Design, and Evaluation. AC/243 (Panel 8) TR/17. North Atlantic Treaty Organization Defence Research Group.

Fallesen, J.J.

1993 Overview of Army Tactical Planning Performance Research. Technical Report 984. U.S. Army Research Institute for Behavioral and Social Sciences.

Fallesen, J.J., C.F. Carter, M.S. Perkins, R.R. Micehl, J.P. Flannagan, and P.E. McKeown

1992 The Effects of Procedural Structure and Computer Support Upon Selecting a Tactical Course of Action. Technical Report 960. U.S. Army Research Institute for Behavioral and Social Sciences.

Fisk, A.D., and F.T. Eggemeier

1988 Application of automatic/controlled processing theory to training tactical command and control skills: I. Background and task analytic methodology. Pp. 1227-1231 in *Proceedings of the Human Factors Society 32nd Annual Meeting*. Santa Monica, CA: Human Factors Society.

Fitts, P.M., and M.I. Posner

1967 *Human Performance*. Westport, CT: Greenwood Press.

Fracker, M.L.

1990 Attention gradients in situation awareness. Pp. 6/1-6/10 in *Situational Awareness in Aerospace Operations AGARD-CP-478*. Neuilly-Sur-Seine, France: Advisory Group for Aerospace Research and Development.

Glance, N.S., and B. Huberman

1993 The outbreak of cooperation. *Journal of Mathematical Sociology* 17(4):281-302.

1994 Social dilemmas and fluid organizations. In *Computational Organization Theory*, K.M. Carley and M.J. Prietula, eds. Hillsdale, NJ: Lawrence Erlbaum Associates.

Grossberg, S., and W.E. Gutowski

1987 Neural dynamics of decision making under risk: Affective balance and cognitive-emotional interactions. *Psychological Review* 94:303-318.

Grossberg, S., and G. Stone (with J. Silverstein, S. Ritz, and R. Jones)

1986 Neural dynamics of word recognition and recall: Additional priming, learning, and resonance. *Psychological Review* 93(1):46-74.

Hartman, B., and G. Secrist

1991 Situation awareness is more than exceptional vision. *Aviation, Space and Environmental Medicine* 62:1084-1091.

Harwood, K., B. Barnett, and C.D. Wickens

1988 Situational awareness: A conceptual and methodological framework. In *Proceedings of the 11th Symposium of Psychology in the Department of Defense*.

Hintzman, D.

1988 Judgments of frequency and recognition memory in a multiple trace memory model. *Psychological Review* 95:528-551.

Holland, J., K. Holyoak, R. Nisbett, and P. Thagard
1986 *Induction: Processes of Inference, Learning, and Discovery.* Cambridge, MA: MIT Press.

Hollenbeck, J.R., D.R. Ilgen, D.J. Sego, J. Hedlund, D.A. Major, and J. Phillips
1995 The multi-level theory of team decision making: Decision performance in teams incorporating distributed expertise. *Journal of Applied Psychology* 80:292-316.

IBM/Army Integrated Development Team
1993 *Close Combat Tactical Trainer (CCTT): Semi-Automated Forces (SAF) Combat Instruction Set (CIS) Development.* Orlando, FL: IBM Integrated Development Team.

John, B.E., and D.E. Kieras
1996 The GOMS family of user interface analysis techniques: Comparison and contrast. *ACM Transactions on Computer-Human Interaction* 3:320-351.

Kaempf, G.L., G.A. Klein, M.L. Thorsden, and S. Wolf
1996 Decision-making in complex command-and-control environments. *Human Factors* 38(2):220-231.

Kahneman, D., P. Slovic, and A. Tversky, eds.
1982 *Judgment Under Uncertainty: Heuristics and Biases.* New York: Cambridge University Press.

Karayiannis, N.B., and A.N. Venetsanopoulos
1993 *Artificial Neural Networks: Learning Algorithms, Performance Evaluation, and Applications.* Boston, MA: Kluwer.

Kephart, J.O., B.A. Huberman, and T. Hogg
1992 Can predictive agents prevent chaos? Pp. 41-55 in *Economics and Cognitive Science*, P. Bourgine and B. Walliser, eds. Oxford: Pergamon Press.

Kieras, D.E., and D.E. Meyer
in press An overview of the EPIC architecture for cognition and performance with application to human-computer interaction. *Human-Computer Interaction.*

Kirwan, B., and L.K. Ainsworth, eds.
1992 *A Guide to Task Analysis.* Bristol, PA: Taylor and Francis.

Klein, G.A.
1994 Naturalistic decision making: Implications for design. *Gateway* 5(1):6-7.

Klein, G.A., R. Calderwood, and A. Clinton-Cirocco
1986 Rapid decision-making on the fire ground. In *Proceedings of the Human Factors Society 30th Annual Meeting.*

Kontopoulos, K.M.
1993 Neural networks as a model of structure. Pp. 243-267 in *The Logics of Social Structure.* New York: Cambridge University Press.

Konz, S.
1995 Predetermined time systems. Pp. 464-475 in *Work Design: Industrial Ergonomics, Fourth Edition.* Columbus, OH: Publishing Horizons, Inc.

Krackhardt, D.
1994 Graph theoretical dimensions of informal organizations. In *Computational Organization Theory*, K.M. Carley and M.J. Prietula, eds. Hillsdale, NJ: Lawrence Erlbaum Associates.

Kruschke, J.K.
1992 ALCOVE: An exemplar based connectionistic model of category learning. *Psychological Review* 99:22-44.

Laird, J.E., P.S. Rosenbloom, and A. Newell
1986 *Universal Subgoaling and Chunking: The Automatic Generation and Learning of Goal Hierarchies*. Boston, MA: Kluwer Academic Publishers.

Laird, J.E., A. Newell, and P.S. Rosenbloom
1987 SOAR: An architecture for general intelligence. *Artificial Intelligence* 33:1-64.

Lant, T.K.
1994 Computer simulations of organizations as experiential learning systems: Implications for organization theory. In *Computational Organization Theory*, K.M. Carley and M.J. Prietula, eds. Hillsdale, NJ: Lawrence Erlbaum Associates.

Lant, T.L., and S.J. Mezias
1990 Managing discontinuous change: A simulation study of organizational learning and entrepreneurship. *Strategic Management Journal* 11:147-179.
1992 An organizational learning model of convergence and reorientation. *Organization Science* 3(1):47-71.

Lawrence, P.R., and J. Lorsch
1967 *Organization and Environment: Managing Differentiation and Integration*. Cambridge, MA: Harvard University Press.

Lin, Z., and K.M. Carley
forthcoming Organizational response: The cost performance tradeoff. *Management Science*.

Lindsey, P.H., and D.A. Norman
1977 *Human Information Processing: An Introduction to Psychology*. Academic Press.

Liu, Y.
1996 Queueing network modeling of elementary mental processes. *Psychological Review* 103:116-136.

Logan, G.D.
1988 Toward an instance theory of automatization. *Psychological Review* 95:492-527.

Macy, M.W.
1991a Learning to cooperate: Stochastic and tacit collusion in social exchange. *American Journal of Sociology* 97(3):808-843.
1991b Chains of cooperation: Threshold effects in collective action. *American Sociological Review* 56:730-747.

Malone, T.W.
1987 Modeling coordination in organizations and markets. *Management Science* 33:1317-1332.

March, J.G., and H.A. Simon
1958 *Organizations*. New York: Wiley.

Masuch, M., and P. LaPotin
1989 Beyond garbage cans: An AI model of organizational choice. *Administrative Science Quarterly* 34:38-67.

McClelland, J.L.
1993 Towards a theory of information processes in graded, random, interactive networks. In *Attention and Performance XIV: Synergies in Experimental Psychology, Artificial Intelligence, and Cognitive Neuroscience*. Cambridge, MA: MIT Press.

McClelland, J.L., and M. Chappell
1994 Bayesian Recognition. Poster given at the 35th Annual Meeting of the Psychonomic Society.

Metcalfe-Eich, J.
1982 A composite holographic associative recall model. *Psychological Review* 89:627-661.

Meyer, D.E., and D.E. Kieras
in press A computational theory of executive cognitive processes and multiple-task performance: Part 1: Basic mechanisms. *Psychological Review*.

Milgram, P., R. van der Wijngaart, H. Veerbeek, O.F. Bleeker, and B.V. Fokker
1984 Multi-crew model analytic assessment of landing performance and decision making demands. In *Proceedings of the 20th Annual Conference on Manual Control*. Moffett Field, CA.

Moses, Y.O., and M. Tennenholtz
1990 Artificial Social Systems. Part I, Basic Principles. Technical Report CS90-12. Weizmann Institute of Science, Department of Computer Science. Makhon Vaitsman le-mada. Rehovot, Israel.

Mulgund, S.S., G.J. Rinkus, C. Ilgen, and G. Zacharias
1996a *A Situation Driven Workload Adaptive Pilot/Vehicle Interface*. Final Report R96051. Cambridge, MA: Charles River Analytics.
1996b A situation driven workload adaptive pilot/vehicle interface. In *Proceedings of the 1996 IEEE Human Interaction with Complex Systems Symposium*. Dayton, OH.

Murdock, B.B., Jr.
1982 A theory for the storage and retrieval of item and associative information. *Psychological Review* 89:609-626.

Myerson, R.B.
1991 *Game Theory: Analysis of Conflict*. Cambridge, MA: Harvard University Press.

National Research Council
1997 *Tactical Display for Soldiers: Human Factors Considerations*. Panel on Human Factors in the Design of Tactical Display Systems for the Individual Soldier, Committee on Human Factors. Washington, DC: National Academy Press.

Newell, A.
1990 *Unified Theories of Cognition*. Cambridge, MA: Harvard University Press.

Nosofsky, R.M., and T.J. Palmari
in press An exemplar based random walk model of speeded classification. *Psychological Review*.

Palmer, J.
1994 Set-size effects in visual search: The effect of attention is independent of the stimulus for simple tasks. *Vision Research* 34(13):1703-1721.

Payne, J.W., J.R. Bettman, and E.J. Johnson
1993 *The Adaptive Decision Maker.* New York: Cambridge University Press.

Pete, A., K.R. Pattipati, and D.L. Kleinman
1993 Distributed detection in teams with partial information: A normative descriptive model. *IEEE Transactions on Systems, Man and Cybernetics* 23:1626-1648.
1994 Optimization of detection networks with multiple event structures. *IEEE Transactions on Automatic Control* 39:1702-1707.

Plaut, D.C., J.L. McClelland, and M.S. Seidenberg
1996 Understanding word reading: Computational principles in quasi-regular domain. *Psychological Review* 103:56-115.

Raaijmakers, J.G.W., and R.M. Shiffrin
1980 SAM: A theory of probabilistic search of associative memory. Pp. 207-262 in *The Psychology of Learning and Motivation: Advances in Research and Theory, Volume 14,* G.H. Bower, ed. New York: Academic Press.
1981 Search of associative memory. *Psychological Review* 88:93-134.

Ratcliff, R., and T. Van Zandt
1995 Comparing Connectionist and Diffusion Models of Reaction Time. Unpublished.

Roberts, K.
1990 Some characteristics of one type of high reliability organizations. *Organization Science* 1(2):160-176.

Schneider, W., and R.M. Shiffrin
1977 Controlled and automatic human information processing: I. Detection, search and attention. *Psychological Review* 84:1-66.

Shiffrin, R.M.
1988 Attention. Pp. 739-811 in *Stevens' Handbook of Experimental Psychology, 2nd Edition,* R.C. Atkinson, R.J. Hernstein, G. Lindzey, and R.D. Luce, eds. New York: Wiley.

Shiffrin, R.M., and G.T. Gardner
1972 Visual processing capacity and attentional control. *Journal of Experimental Psychology* 93:72-82.

Shiffrin, R.M., and W. Schneider
1977 Controlled and automatic human information processing: II. Perceptual learning, automatic attending, and a general theory. *Psychological Review* 84:127-190.

Shiffrin, R.M., and M. Steyvers
in press A model for recognition memory: REM: Retrieving effectively from memory. *Psychological Bulletin and Review* (June 1997).

Simon, H.
1955 Behavioral model of rational choice. *Quarterly Journal of Economics* 69:99-118.
1956 Rational choice and the structure of the environment. *Psychological Review* 63:129-138.

Sperling, G., and B.A. Dosher
1986 Strategy and optimization in human information processing. Pp. 2-1–2-65 in *Handbook of Perception and Human Performance: Volume 1.* New York: Wiley.

Sperling, G., and E. Weichselgartner
1995 Episodic theory of the dynamics of spatial attention. *Psychological Review* 102(3):503-532.

Spick, M.
1988 *The Ace Factor*. Annapolis, MD: Naval Institute Press.

Stiffler, D.R.
1988 Graduate level situation awareness. *USAF Fighter Weapons Review* Summer: 115-120.

Tambe, M.
1996 Tracking dynamic team activity. In *Proceedings of the National Conference on Artificial Intelligence (AAAI)*.

Tambe, M., W.L. Johnson, R.M. Jones, F. Koss, J.E. Laird, P.S. Rosenbloom, and K. Schwamb
1995 Intelligent agents for interactive simulation environments. *AI Magazine* 16(1).

Tang, Z., K.R. Pattipati, and D.L. Kleinman
1993 Optimization of detection networks: Part II—Generalized tree structures. *IEEE Transaction on Systems, Man and Cybernetics* 23:211-221.

Townsend, J.T., and F.G. Ashby
1983 *The Stochastic Modeling of Elementary Psychological Processes*. New York: Cambridge University Press.

Townsend, J.T., and J.R. Busemeyer
1995 Dynamic representation of decision making. Pp. 101-120 in *Mind as Motion*, R. Port and T. van Gelder, eds. Cambridge, MA: MIT Press.

U.S. Department of the Army
1993 *FM 100-5 Operations*. Headquarters. Fort Monroe, VA: U.S. Department of the Army.

U.S. Department of Defense
1995 *Modeling and Simulation (M&S) Master Plan*. Training and Doctrine Command. October. Washington, DC: Department of Defense.
1996 *Verification, Validation and Accreditation (VV&A) Recommended Practices Guide*. Office of the Director of Defense Research and Engineering. Washington, DC: Defense Modeling and Simulation Office.

Wickens, C.D.
1984 Processing resources in attention. In *Varieties of Attention*, R. Parasuraman and D.R. Davies, eds. Orlando, FL: Academic Press.

Zacharias, G., A. Miao, C. Ilgen, J. Yara, and G. Siouris
1996 SAMPLE: Situation Awareness Model for Pilot-in-the-Loop Evaluation. First Annual Conference on Situational Awareness in the Tactical Air Environment, Naval Air Warfare Center, Patuxent River, MD.

Zacharias, G.L, A. Miao, A. Kalkan, and S. Kao
1994 Operator-based metric for nuclear operations automation assessment. In *Transactions of the Twenty-Second Water Reactor Safety Information Meeting*. Bethesda, MD.

Zacharias, G.L., A.X. Miao, E.W. Riley, and R.K. Osgood
1992 Situation Awareness Metric for Cockpit Configuration Evaluation. Final Report AL-TR-1993-0042. Armstrong Laboratory. Wright-Patterson Air Force Base, OH.